INTEGRATED SMART SENSORS

INTEGRATED SMART SENSORS
DESIGN AND CALIBRATION

by

Gert van der Horn
University of Delft

and

Johan L. Huijsing
University of Delft

SPRINGER-SCIENCE+BUSINESS MEDIA, B.V.

A C.I.P. Catalogue record for this book is available from the Library of Congress.

ISBN 978-1-4419-5016-1 ISBN 978-1-4757-2890-3 (eBook)
DOI 10.1007/978-1-4757-2890-3

Printed on acid-free paper

Table of Contents

1 Introduction 1

1.1 Introduction 1
1.2 Sensors and actuators 2
1.3 Integrated smart sensors 5
1.4 Measurement errors and correction 6
1.4.1 Process variations and time variations 7
1.4.2 Calibration 8
1.4.3 Self-testing 11
1.4.4 Self-calibration 13
1.4.5 Redundancy improvement or feedforward testing 15
1.4.6 Concluding remarks 16
1.5 Objective and Organization 16
1.5.1 Objective 16
1.5.2 Organization 17
1.5.3 Design constraints 17
References 19

2 Integrated Smart Sensor Concept 21

2.1 Introduction 21
2.2 Silicon sensors 22
2.2.1 Temperature sensor 22
2.2.2 Pressure sensor 23
2.2.3 Measurement errors 25

2.3 Analog Interface circuits 27
2.3.1 Temperature sensor readout 27
2.3.2 Pressure sensor readout 32
2.4 Analog-to-Digital conversion 33
2.4.1 Sigma-delta converter 34
2.5 Digital bus or microcontroller interface 39
2.5.1 Microcontroller compatible interfaces 39
2.5.2 Smart Sensor bus interfaces 40
2.6 Integrating a calibration function in the smart sensor concept 43
2.6.1 Error sources in the smart sensor signal path 43
2.6.2 Conventional sensor calibration techniques 44
2.6.3 Correction of the sensor signal transfer 45
2.6.4 Calibration memory 46
References 47

3 *Calibration and Linearization Techniques* *51*

3.1 Introduction 51
3.1.1 Errors in the sensor transfer curve 51
3.1.2 Systematic and random errors 53
3.1.3 Linear and nonlinear errors 55
3.1.4 Normalized sensor transfer functions 56
3.2 Linearization 58
3.2.1 Sensor characteristic linearization 59
3.2.2 Linearization based on a look-up table 60
3.2.3 Linearization based on piecewise-linear interpolation 63
3.2.4 Linearization based on piecewise-polynomial or spline interpolation 67
3.2.5 Linearization based on curve fitting 70
3.2.6 Linearization based on error minimization 73
3.2.7 Conclusion 75

3.3 Progressive polynomial calibration method 76
3.3.1 One-dimensional functions 77
3.3.2 Two-dimensional functions 84
3.3.3 Multi-dimensional functions 88
3.4 Conclusion 89
References 90

4 *Calibration using Analog Signal Processing* 93

4.1 Introduction 93
4.2 Conventional sensor calibration 94
4.2.1 Resistor trimming 94
4.2.2 Programmable resistor-arrays 96
4.3 Analog calibration circuits 96
4.3.1 Classical offset and full-scale calibration 98
4.3.2 Classical cross-sensitivity calibration 99
4.3.3 Improved analog calibration circuit 102
4.4 Classical pulse-modulated calibration 104
4.5 Analog implementation of a polynomial calibration 106
4.5.1 System configuration 106
4.5.2 Analog current multiplier 108
4.5.3 Programmable current divider DAC 111
4.5.4 VI-converters and IV-converters 113
4.5.5 System implementation 116
4.5.6 System operation 117
4.5.7 Realization 121
4.6 Conclusion 124
References 126

5 *Calibration using Sigma-Delta Analog-to-Digital Conversion* 127

5.1 Introduction 127

5.2 Sigma-delta AD-converters 128

5.2.1 Principle of oversampling 129
5.2.2 Principle of noise shaping 130
5.2.3 Principle of decimation 136
5.2.4 Transfer from analog input to digital output 139

5.3 Alteration of the sigma-delta transfer 140

5.3.1 Digital sigma-delta bitstream generator 140
5.3.2 Modulating the input signal for gain calibration 143
5.3.3 Modulating an additional signal for offset calibration 147
5.3.4 Sigma-delta calibration configurations 148

5.4 Implementation of the sigma-delta AD-converter 152

5.4.1 Differential integrator amplifier 153
5.4.2 Clocked comparator 155
5.4.3 Sigma-delta converter 157
5.4.4 Digital decimation filter implementation 159

5.5 Smart temperature sensor realizations 160

5.5.1 Smart temperature sensor with calibration and IS^2 bus interface 161
5.5.2 Temperature sensor with linearity calibration 165

5.6 Conclusion 168

References 169

6 Calibration using Digital Signal Processing 173

6.1 Introduction 173

6.2 Hardware implementation 174

6.3 Software implementation 178

6.3.1 Sensor system configuration 178
6.3.2 Calibration algorithm 179
6.3.3 Calibration program 189
6.3.4 Future work 191

6.4 Conclusion 191

References 192

7 Summary and Conclusions 193

Index 199

Introduction

1

1.1 Introduction

The (signal processing and storage) capacity of the human brain enables us to become powerful autonomous beings, but only if our brains operate in conjunction with (at least some of) our senses and muscles. Using these organs, we can interact with our environment, learn to adapt, and improve important aspects of our life. Similarly, the signal processing capabilities of modern electronics (computers) could be combined with electronic sensors and actuators to enable interaction with, and adaptation to, the (non-electrical) environment. This will lead to smarter and more powerful automated tools and machines. To facilitate and stimulate such a development, *easy-to-use low-cost* sensors are needed. The combination of electronic interface functions and a sensor in an *integrated smart sensor,* that provides a standard, digital, and bus-compatible output, would simplify the connection of sensors to standard electronic signal processors (microcontrollers, computers, etc.). Currently, the calibration procedure, required for standardization of the sensor output signal level, contributes largely to the production costs of accurate sensors. To enable automation of the calibration procedure, and hence reduce the sensor fabrication costs, a *digital calibration function* should be included in the smart sensor.

1.2 Sensors and actuators

In industry many processes are electronically controlled. As depicted in Fig. 1-1, to be in control of a process one needs to be able to gather all necessary information about that process through sensors, then to examine that information and maybe modify it using a controller, and take the proper actions through actuators to influence the process according to the desired algorithm [1]. The results of these actions can be measured, which enables a feedback control [2].

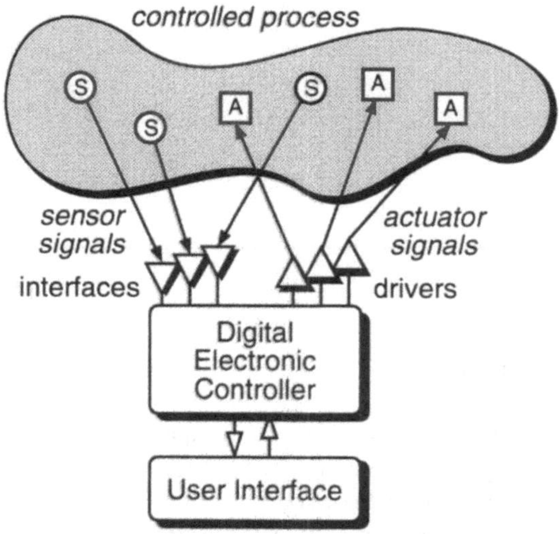

Fig. 1-1 *Gathering the process information, and controlling the process, using sensors and actuators.*

To gather process information we measure process parameters, such as pressure and temperature, using sensors. To control the pressure and temperature of a process we use actuators, such as valves and heaters. Due to the evolution in microelectronics of the past decades we now have powerful control and processing tools available which operate in the electrical signal domain. Therefore, we need sensors which convert signals from any of the five other signal domains; mechanical, thermal, chemical, magnetic, and optical to an electrical signal which can be processed electronically and passed on to an advanced digital computer [7]. For the control action, we need actuators which can be electrically driven to effect changes in one or more of the five other signal domains.

Based on a large diversity of underlying physical principles, many sensors and actuators have been and are being developed. Each of these devices has its own unique properties and requires its own special electronic interface. Therefore, the realization of a system as depicted in Fig. 1-1 has become the terrain of many different technical specialists, thus limiting the possibilities from a system-level point of view.

From a system-level point of view, a configuration as shown in Fig. 1-2 is much more practical: sensors and actuators that are able to communicate with a digital controller using standardized signals on a low-level communication bus. The digital controller is able to communicate signals on a high-level bus to a user interface, for example.

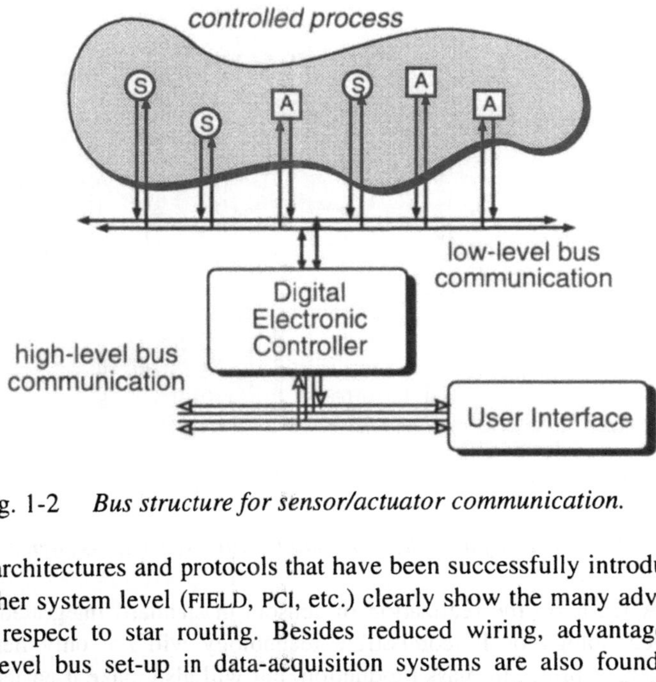

Fig. 1-2 *Bus structure for sensor/actuator communication.*

Bus architectures and protocols that have been successfully introduced on a higher system level (FIELD, PCI, etc.) clearly show the many advantages with respect to star routing. Besides reduced wiring, advantages of a low-level bus set-up in data-acquisition systems are also found in the exchangeability of devices, in multiplexing options, and in easy maintenance and expansion [3,4,5]. Furthermore, it simplifies division of a complex system with high-level bus communication, into sub-systems using low-level bus communication; Fig. 1-3 shows how the system can be divided into a data-acquisition sub-system to interface the sensors, and another sub-system for controlling the actuators, each having it's own low-level bus and digital controller. It is also possible that some devices communicate on a high-level [6] and others on a low-level bus [1].

To enable the proposed configurations, many interfacing functions will have to be integrated with the sensor or actuator device. Such *smart sensors* and *smart actuators* have been only sparsely available up to now. Whereas most actuators, micro-actuators excluded, are large and power consuming, the addition of smart electronic circuits to generate driving signals will enhance performance, for a relatively small increase in price and energy. Sensors are generally micro-devices which are often manufactured using special processing. At first sight, the investment needed to include or add electronic functions with a sensor device seems relatively large.

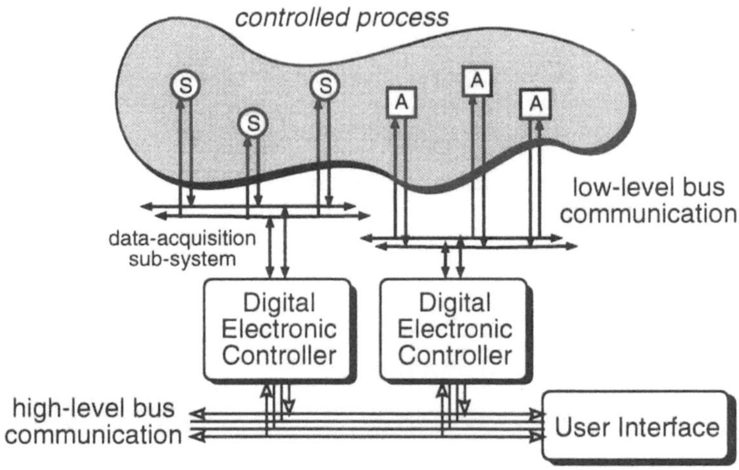

Fig. 1-3 *System division into two low-level sub-systems.*

A focus on the realization of sensors in standard Integrated Circuit (IC) technology or IC-compatible technology will not only help to reduce sensor prices in mass production, but will also make it easier to integrate the required functions in electronic circuits with a sensor device [8,9,10]. In the next paragraph we will focus on the functions that should be integrated on a smart sensor, in order to simplify setting up configurations as proposed in Fig. 1-2 and Fig. 1-3.

1.3 Integrated smart sensors

In order to realize a smart sensor which is able to communicate on a standard bus level as previously proposed, the following functions must be included in the sensor package [8]:

- *sensor(s), preferably silicon sensor(s)*
- *analog sensor readout and signal conditioning*
- *analog-to-digital (AD) conversion, to provide a digital output signal*
- *bus interface, to simplify communication to microcontrollers, PCs, and other devices*
- *calibration of the sensor transfer curve, preferably a digitally programmable calibration*

First of all, the sensor converts the physical input signal into an electrical signal. Second, the usually weak analog sensor signal needs to be amplified and converted into convenient electrical signals (voltage or current, rather than capacitance, resistance, etc.). Third, an AD-converter is necessary to allow communication with the digital world of microcontrollers and computers, often a serial 1-bit output is preferred here. To simplify setting up systems with multiple sensors and actuators, each device will also need a bus interface on board to transport the sensor signal to a digital microcontroller or computer, which further processes the sensor signal into a digital multi-bit (e.g. bytes) format. Fig. 1-4 shows the signal processing chain of a smart sensor containing these functions, and clarifies the signal conversion realized in each block.

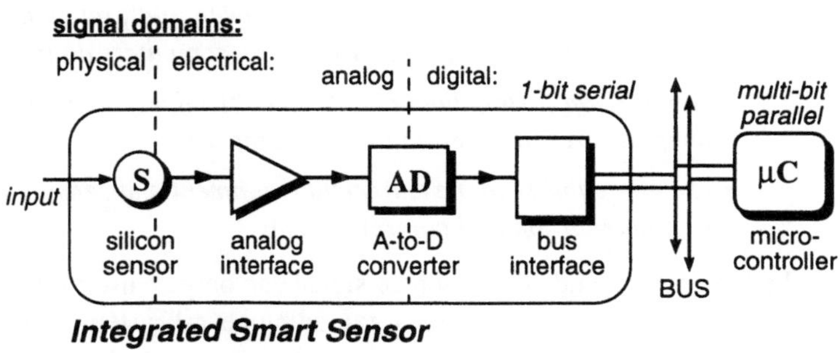

Fig. 1-4 *Signal processing chain of an integrated smart sensor.*

Last but not least, on-board calibration electronics will make it possible to realize smart sensors with standardized output signals, allowing easy exchange of sensors and easy re-calibration [8,11]. Calibration is necessary to 'pinpoint' the transfer from the physical input signal, temperature for example, to the electrical output signal, so that the manufacturer can guarantee that a temperature of 0 degrees Celsius always corresponds to a digital output signal 0. In the manufacturing of the sensors this transfer often shows a large device-to-device variation. Different options to correct this will be discussed in the next paragraph.

1.4 Measurement errors and correction

Before studying measurement errors in sensor systems and ways to correct such errors, it is important to contemplate what we want to measure and what we intend to do with the information on the measurand. It is not always necessary to relate the sensor signal to a well-defined standard unit (SI[1] or SI-derived unit). In a trained (expert) or self-learning (neural) system, a sensor signal is judged *subjectively*, based on experience, to take the proper actions; our human bodies are not equipped to measure the temperature outside in degrees Fahrenheit, Celsius or in Kelvin, yet with our senses we are perfectly capable to make a healthy decision on whether to go outside in shorts or to put on a coat. The scientists named have taught us that we can relate our human temperature sensing to standardized numbers, which we can communicate in weather reports for example. This brings us to the advantages of relating the sensor signal to the measurand in a standardized unit, such as Kelvin for temperature:

- *interpretation; we can easily learn how to interpret numeral information and, moreover, we can let computers do it*
- *comparison; using numbers we can objectively compare temperatures at different times and places*
- *exchangeability of measurement equipment and sensors in instrumentation and control systems*

We have to conclude that a sensor signal can only be usefully processed in measurement or control systems, excluding neural systems, when it can be related to a well-defined standard unit. In some control systems the target

[1] SI=Système International d'unités, or International System of units.

is to regulate an error signal to zero. In that case it is sufficient to know only the offset, while non-zero signals do not have to be known in precise units. For example, when placing a robot arm on top of an object it can be sufficient to know when the measured distance between the arm and object has become zero.

The process of relating the measurement or sensor signal to the physical input signal in precise well-defined units is referred to as *calibration.*

1.4.1 Process variations and time variations

Calibration would be simple if all sensor devices in a production line had the same properties. In that case, the relation between electrical output signal and physical input signal should be traced (measured) for only one sensor, and then it can be applied to all sensors of that type. Unfortunately, many variations in the production process must be tolerated and, therefore, also the properties of the sensors will vary from device to device. Particularly in IC-technology we know to expect large absolute variations (50%) and smaller relative (on-chip) variations (1%). For example, micromachined pressure sensors based on piezoresistors show large variations in offset (due to resistor mismatch, initial stress, etc.), in sensitivity (position/orientation of resistors, mechanical properties of the diaphragm, etc.), and also in cross-sensitivity to temperature. An error band in the output signal range as large as 30% of the full-scale can be expected [11]. Each sensor device will have to be calibrated individually in order to guarantee that it exhibits a well-defined and fixed transfer from input to output. This usually means that each device must be measured individually, in a controlled environment, using reference signal conditions.

This individual attention per device is very expensive in mass production, which explains the large price difference between uncalibrated mass-produced sensors and calibrated measurement equipment. To reduce the costs of the sensor device calibration, we should look for ways to automate the calibration of many sensors at a time, in a so-called batch calibration. The smart sensor configuration proposed in the previous section makes this possible: many such smart sensors can be put (simultaneously) in a controlled environment (such as a temperature oven or a pressure chamber), via a bus connection each sensor output can be measured quickly, and the integrated calibration can be programmed to obtain precisely the desired transfer. Furthermore, the number of necessary reference measurements must be minimized, in order to save time and thus fabrication costs.

Another problem which can be encountered in the transfer of a sensor is that it may alter slowly over time; some sensor characteristics depend on material properties which degrade or "wear off". Although these changes are usually small, over a long period (months or years) they may cause a significant error, known as *drift*. This drift error has to be reset at a certain interval by a *re*-calibration. For this reason, it often makes sense to integrate at the sensor some kind of actuator, which can generate a well-defined, additional input signal for the sensor; this introduces the possibility of self-test and self-calibration. The terms calibration, self-test, and self-calibration in the context of sensor technology[1] will be further clarified in the succeeding paragraphs.

1.4.2 Calibration

To define calibration in general we quote Nicholas & White [12]:

> "The word 'calibration' has changed its meaning with time and three distinct meanings can be distinguished. ...
>
> The word appears to derive from an Arab word for a mould for casting metal. This is presumably because calibration originally referred to the making of guns by casting metal. Calibration referred either to the means used to determine the *calibre* (or bore) of the gun or to the determination of the range of the gun. ...
>
> The second and more common meaning of the word calibration is the checking and adjusting of the bore of an instrument and the application of a *scale* by the manufacturer. ... That is, a calibration refers to the set of operations carried out by an instrument manufacturer in order to *ensure* that the equipment has a *useful* measurement scale. ..."

It is difficult to objectively define the terms *scale*, *ensuring*, and *usefulness*, so a more official definition of the term calibration is unavoidable:

[1] Unfortunately the terms self-test and auto-calibration are also popular in electronic circuit technology, especially for sensor interface circuits. The terms could be used with much more care, by specifying exactly what is tested or automatically adjusted.

> "The third meaning for calibration, as used by metrologists, is defined by the ISO (International Organization for Standardization) as:
>
> CALIBRATION, *The set of operations which establish, under special conditions, the relationship between values indicated by a measuring instrument or measuring system, or values, represented by a material measure, and the corresponding known value of a measurand.*
>
> NOTES, (1) *The result of a calibration permits the estimation of errors of indication of the measuring instrument ... or the assignment of values to marks on arbitrary scales.* (2) *A calibration may also determine other metrological properties.* (3) *The result of a calibration may be recorded in a document, sometimes called a calibration certificate or a calibration report.* (4) *The result of a calibration is sometimes expressed as a calibration factor, or as a series of calibration factors in the form of a calibration curve.*
>
> This definition is best seen as a pointer to a meaning of calibration acceptable to modern quality assurance and accreditation practices. It represents the current practice and if emphasis is put on the NOTES the definition gives rise to a valuable concept of calibration. ..."

The official side of measurement standards and calibration is the concern of national and international standardization institutes and will not be an issue in this book.

In practice, a sensor calibration is performed by putting the sensor in a controlled environment, as indicated by *phase 1* in Fig. 1-5. The environment is controlled in such a way that physical reference signals can be generated in accordance with the official standards. A temperature sensor can be inserted in an oven, in which a heater system can create a sensor ambience with a temperature of 300, 350 or 400 Kelvin, for example. Usually an extra (officially calibrated) reference sensor or instrument is used to guarantee these temperatures. Sometimes, more (reference) sensors are needed to monitor other ambient parameters, such as humidity. The measurements of the sensor output signal at several temperatures can be used to determine the error with respect to the desired sensor transfer. Depending on the sensor characteristics and the number of reference measurements the sensor transfer and error curve can be determined within a certain accuracy. This is not sufficient; the calibration needs to be completed by adjusting or correcting the sensor transfer with trimmable (adjustable) elements in such a way that an accurately specified transfer is obtained, as illustrated in Fig. 1-6.

At a set of input signals $[x_1 ... x_n]$, which can be calculated by using the known transfer of the reference sensor, the sensor outputs $[y_1 ... y_n]$ are measured. By using a correction table or a correction function, the sensor

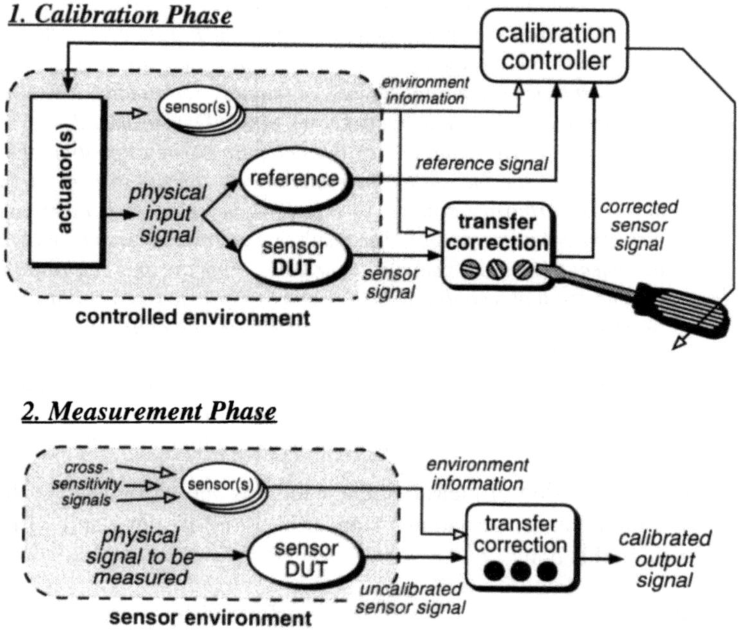

Fig. 1-5 *Schematic explanation of sensor calibration.*

outputs are processed in order to retrieve a signal which relates to the original physical signal in accordance with the desired transfer curve.

After calibration the correction should be maintained so that the calibrated sensor can be used to reliably measure the physical signal in an uncontrolled environment, as indicated by *phase 2* in Fig. 1-5. The manner in which the correction curve is obtained and implemented can be referred to as the *calibration method.* A traditional calibration method for pressure sensors is to insert trimmable resistors (potentiometers or laser-trimmed) in the Wheatstone bridge and manually adjust them, see also section 4.2.1. This book will discuss several possibilities of implementing a more attractive *digitally-programmable transfer correction* in the circuitry of a smart sensor interface.

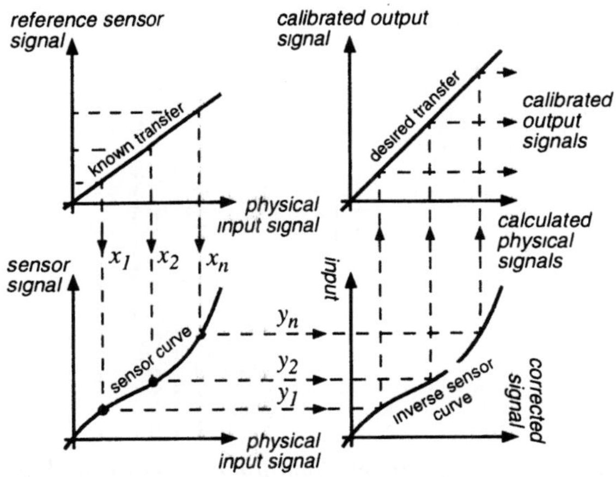

Fig. 1-6 *Determining the sensor transfer and calibration curve.*

To summarize, the essential elements of calibration are listed as:

- *a calibration phase in a controlled environment*
- *generation of physical reference signals for sensor excitation, and measurement of the sensor output signal*
- *correction of the signal transfer based on those calibration measurements*

Officially, the reference signals used in the calibration measurements should be traceable[1] to an SI-unit.

When referring to a 'sensor with calibration' we mean that the possibility of correcting the sensor transfer is integrated with the sensor or sensor interface.

1.4.3 Self-testing

By installing one or more actuators close to the sensor, see Fig. 1-7, it is possible to influence the physical input signal of the sensor at the command of an on-chip control circuit. This controller then verifies whether the sensor has detected the change in signal, if not it generates an error signal to indicate that the sensor (or actuator) is not functional. This

[1] Traceability is also clearly defined in Nicholas &White [12]

can avoid incorrect interpretation of the sensor signal inside the system in which the sensor is used.

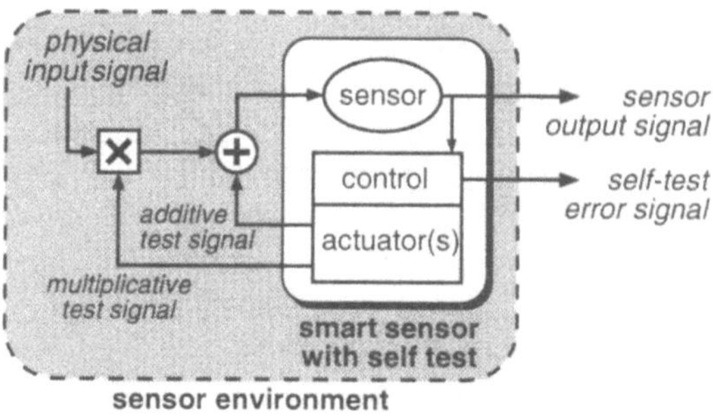

Fig. 1-7 *Schematic explanation of a sensor with self-test.*

For some types of sensors we can realize an additive, for others a multiplicative test signal. On a temperature sensor, for example, we can integrate a heater element (i.e. a resistor) which, when turned on, increases the chip temperature with respect to the ambient. An additive test signal of a few degrees can be realized. It is important to realize that the self-test signal has to be detectable under all conditions, since the sensor environment is not controlled as in the case of calibration. During the self-test the sensor output signal should be ignored.

Although from the operation point of view we might have improved reliability, from the production point of view we have added an extra element which can fail, namely the actuator. This may affect the yield in the production and the mean-time-before-failure of the product. It is thus important to verify that the reliability of the actuator is not much worse than that of the sensor. For many types of sensors this will not be the case, or an appropriate actuator which can be integrated with the sensor will not be available at all.

Summarizing the essential elements of a sensor with a self-test function:

- *an integrated actuator which*
- *allows a test of the functionality*
- *realizes an improved reliability*
- *operates in an uncontrolled normal sensor environment*

1.4.4 Self-calibration

As can be seen in Fig. 1-8, the self-calibration[1] function is very similar to the self-test function, but needs the addition of accurate actuator(s) and a possibility of correcting the transfer, based on sensor measurements which can be obtained by using different actuator signals.

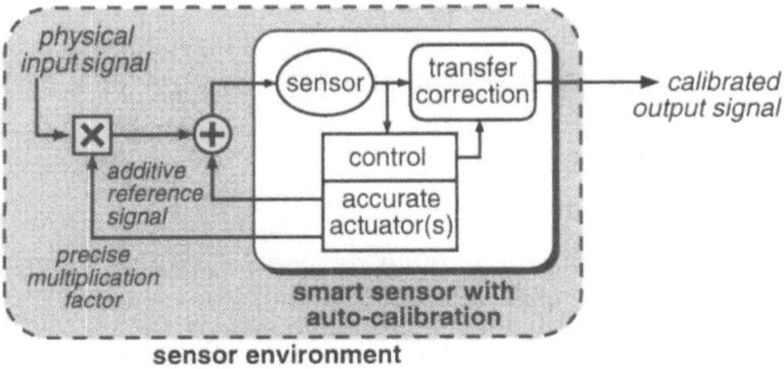

Fig. 1-8 *Schematic explanation of a sensor with self-calibration.*

It is essential for a self-calibration function that the actuators can be used to improve the properties of the sensor, which means that the transfer of the actuator has to be more accurate, in some respects, than the transfer of the sensor. Few examples of sensor/actuator combinations for which this is true can be found. One example is that of a magnetic Hall-plate sensor which may exhibit a relatively large drift in sensitivity and also a cross-sensitivity to temperature [13]. As a magnetic actuator, a coil can be realized around the sensor. With an accurate reference current and a well-defined geometry, the coil produces an accurate magnetic signal which is added to the magnetic input signal. The resulting change in sensor output can be measured and should always be the same (for a constant sensitivity). To correct the measured sensitivity errors, the current through the Hall plate can be adjusted. It is also necessary to be able to separate the self-calibration signal from the physical input signal using a multiplexing technique, either in time or frequency.

[1] Often the term auto-calibration is used, which means the same as self-calibration if the word auto is interpreted as the Greek word for self, as in autodidact, but not as an abbreviation for automatic.

To self-calibrate for offset errors, a multiplicative actuation signal is needed. The magnetic sensor, for example, could be mechanically turned around 180 degrees, or electronically chopped by inverting the Hall-plate current [14]. This should result in an exactly opposite sensor output signal while the offset stays the same. In fact, the signal can be multiplied by plus or minus one. The offset is found by taking the average of the two (inverted and non-inverted) signals. By subtracting the two signals the offset is eliminated.

Essentials for self-calibration, in addition to those for self-testing, are:

- *accurate actuators*
- *transfer correction*
- *clear separation of input signal and calibration signal*

Since it is very difficult, if not impossible, to realize on-board actuators which generate independent reference signals traceable to SI-units, officially the term self-calibration can be considered contradictory [12]. Self-calibration of sensors does not eliminate the need for 'real' calibration. It should rather be seen as a method to monitor the parameters of a sensor and *automatically adjust* them, so that the time between (re-)calibrations can be extended. More precisely, we could speak of *auto-zero adjust* and *auto-gain* or *auto-scale adjust* instead of self-calibration.

It is useful to extend the potential of self-test and self-calibration by considering tandem transducers. A tandem sensor obtains the desired transfer from physical input signal indirectly via another physical signal into the electrical domain, and vice versa for a tandem actuator [7]. In such cases we can distinguish self-test or self-calibration options on different levels. For example, in the case of a thermal flow sensor, which first converts the flow signal to a thermal signal and then to an electrical signal, it will be difficult to realize a self-test signal for flow, whereas thermally, using a heater, it can be feasible. In the case of a capacitive accelerometer it is possible to use a tandem actuation for generating a self-test signal: by applying a charge on the sensor mass and generating an electrical field, an electrostatic force and thus an acceleration signal result [15,16].

1.4.5 Redundancy improvement or feedforward testing

In the self-test and self-calibration principles explained above, an actuator was used to enable *feedback* to the sensor input signal. In the feedback loop the sensor and actuator operate in *series*, which means that the ensemble can only work properly if both the sensor and the actuator are operational. Hence the smart sensor can test itself. If the actuator and sensor have exactly compensating errors this may remain undetected, but this is very unlikely.

It is also possible to improve reliability by putting multiple sensors in *parallel* or by using extra sensors to monitor and *feedforward* the operating conditions, as suggested in Fig. 1-9. By processing the signals of

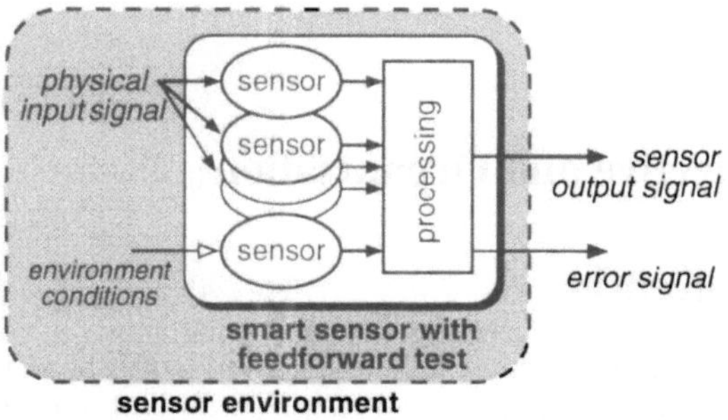

Fig. 1-9 *Schematic explanation of feedforward self-test using multiple sensors.*

multiple sensors in a certain way, single sensor defects can be detected and failure of the measurement can be prevented. Defects or "out-of-range" indications can be indicated with an error signal. By using the increased redundancy provided in the combination of the sensor signals, a more reliable measurement of the physical input signal is realized. The reliability is best improved when combining dissimilar sensors with different failure mechanisms.

It is important to realize that the feedforward principle does not make it possible to test the *operation* of the sensor but only to verify the *validity* of the output signal.

1.4.6 Concluding remarks

For many sensor types self-test or self-calibration is not (yet) feasible because the essential actuator does not exist or does not fulfil the requirements. Unless an actuator with a standardized reference signal can be integrated, self-calibration cannot eliminate the need for calibration[1], as described in section 1.4.2. Only in neural (learning) or fuzzy controlled systems the relation of the sensor signal to a precisely defined unit is not so important. In all other cases, sensor calibration is indispensable. Whenever the sensor output signal must relate precisely to an agreed standard unit for the measured physical signal, it is necessary to correct the sensor transfer based on calibration measurements. This is unavoidable if we want to make sensors a 'plug-and-play' technology.

1.5 Objective and Organization

1.5.1 Objective

The objective of the research discussed in this book is the incorporation of a calibration function in the smart sensor concept. The integration of such a calibration function with the sensor device, next to other smart sensor functions such as AD-conversion and bus interface, is an attempt to develop accurate, economically feasible sensors. It is important that the calibration function can be programmed digitally because this will allow automation of the manufacturer's calibration procedure which, in conjunction with batch calibration, can minimize the calibration costs per sensor. Once reliable smart sensors are available at low cost they will not only be applied in industry, but they may also enhance consumer products and improve our personal comfort.

[1] Nevertheless, the development of self-test and self-calibration techniques should be further stimulated, as it offers other advantages, especially with respect to reliability improvement of the sensor.

1.5.2 Organization

By recalling the smart sensor signal chain presented in Fig. 1-4 we can clarify the organization of this book:

In *chapter 2* the 'state-of-the-art' *smart sensor* concept will be explained, each of the *function blocks* - sensor, analog interface, AD-converter, and bus interface - will be discussed briefly and illustrated by implementation examples.

In *chapter 3* the errors which may be encountered in the transfer function of a sensor will be explained. Different approaches to linearize the sensor transfer function by means of calibration will be explained from a theoretical point of view. A special polynomial calibration method will be presented.

The practical possibilities to correct the overall signal transfer of a smart sensor for calibration purposes, can roughly be divided in three categories, according to the three function blocks in which the correction can be implemented. These categories are issued in the subsequent chapters.

In *chapter 4* different options to change the transfer of analog circuits in a digitally programmable way are examined. The one-dimensional polynomial calibration method presented in chapter 3 is worked out in the realization of a programmable analog calibration circuit.

In *chapter 5* digital alteration of the transfer of the sigma-delta AD-converter by use of bitstream signals is discussed. This involves *mixed-mode* circuits. It is illustrated by the realization of a digitally programmable smart temperature sensor.

In *chapter 6* the proposed polynomial calibration method is extended to the two-dimensional case, considering pressure sensors which have a cross-sensitivity to temperature. The calibration algorithm is implemented in a simple microcontroller program, which realizes the calibration correction in the *digital* signal domain.

1.5.3 Design constraints

Considering the complications of designing modern smart sensor interface circuits (certainly as single-designer projects), it was decided to make use of only one IC process, and hence reduce the effort required to set up the

CAD tools for the IC design flow. The 3-Ghz bipolar process 'Dimes-01' of the Delft Institute for Microelectronics and Submicron-technology, DIMES, was available at low cost, and was therefore an obvious choice for this research project.

Although the choice for the Dimes-01 process may not be optimal from the economical point of view for all the encountered designs, a consistent use of the same process throughout the research project was favored. Furthermore, the many silicon sensor developments in the Dimes-01 process can now easily profit from the advances in the sensor interface designs in the same process.

References

[1] F.R. Riedijk, *Integrated Smart Sensors with Digital Bus Interface*, PhD Thesis, Delft University Press, Delft, 1993.

[2] R.C. Dorf, *Modern Control Systems, Fourth Edition*, Addison-Wesley Publishing Company, Reading, Massachusetts, 1986.

[3] W.A. Koeleman and P.P.L. Regtien, "Bus-organized data-acquisition systems", *Sensors and Actuators*, Vol.5, 1984, pp.327-333.

[4] J.E. Brignell, "Sensors in distributed instrumentation systems, *Sensors and Actuators*, Vol.10, 1986, pp.239-248.

[5] N. Najafi and K.D. Wise, "An organization and interface for sensor-driven semiconductor process control systems", *IEEE Trans. Semiconductor Manufacturing*, Vol.3, No.4, November 1990, pp.230-238.

[6] M. Rutka, *Integrated Sensor Bus*, PhD Thesis, Delft University Press, Delft, 1994.

[7] S. Middelhoek and S.A. Audet, *Silicon Sensors*, Academic Press, London, 1989.

[8] J.H. Huijsing, F.R. Riedijk, and G. v.d. Horn, "Developments in integrated smart sensors", *Sensors and Actuators A*, Vol.43, 1994, pp.276-288.

[9] R.F. Wolffenbuttel, edit., *Silicon Sensors and Circuits: on-chip compatibility*, Chapman & Hall, London, 1996.

[10] R.F. Wolffenbuttel, "Fabrication compatibility of integrated silicon smart physical sensors, *Sensors and Actuators A*, Vol.41-42, 1994, pp.11-28

[11] J. Bryzek, K. Petersen, J.R. Mallon, L. Christel, and F. Pourahmadi, *Silicon Sensors and Microstructures*, NovaSensor, Fremont, USA, 1990.

[12] J.V. Nicholas and D.R. White, *Traceable Temperatures: an introduction to temperature measurement and calibration*, John Wiley &Sons, Chichester/NewYork, 1994.

[13] P.L.C. Simon, P.H.S. de Vries, and S. Middelhoek, "Autocalibration of silicon Hall devices", *Sensors and Actuators A*, Vol.52, 1996, pp.203-207.

[14] M.J.A.M. van Putten, M.H.P.M. van Putten, and A.F.P. van Putten, "Full additive drift elimination in vector sensors using the alternating direction method (ADM)", *Sensors and Actuators A*, Vol.44, 1994, pp.13-17.

[15] R.P. van Kampen, *Bulk-Micromachined Capacitive Servo-Accelerometer*, PhD Thesis, Delft University Press, Delft, 1995.

[16] H.V. Allen, S.T. Terry, and D.W. de Bruin, "Accelerometer systems with built-in testing", *Sensors and Actuators A*, Vol.21-23, 1990, pp.381-386.

Integrated Smart Sensor Concept

2

2.1 Introduction

In this chapter we will present a concept for implementing smart sensor interfaces [1]. Within the scope of this work we cannot discuss many types of sensors or sensor interfaces, hence we limit ourselves to two common sensor types, for measuring temperature, and for measuring pressure. This limitation does not mean that the presented matter cannot be extended to other types of sensors.

We will explain the basic principle of operation of a monolithic temperature sensor, and a pressure sensor based on piezoresistors. We will then show the circuitry which is necessary to read out the sensors and derive signals suitable for AD-conversion. The preferred type of AD-converter, a sigma-delta converter, will also be explained, as well as a possible serial bus interface.

2.2 Silicon sensors

The temperature sensor and the pressure sensor are both IC-compatible and thus very suitable examples in the context of smart sensors.

2.2.1 Temperature sensor

The difference in base-emitter voltage of two transistors with different current densities exhibits a linear relation with respect to the absolute temperature [2,3]. We take two equal transistors and bias them with currents that are scaled by a factor of four, as illustrated in Fig. 2-1.

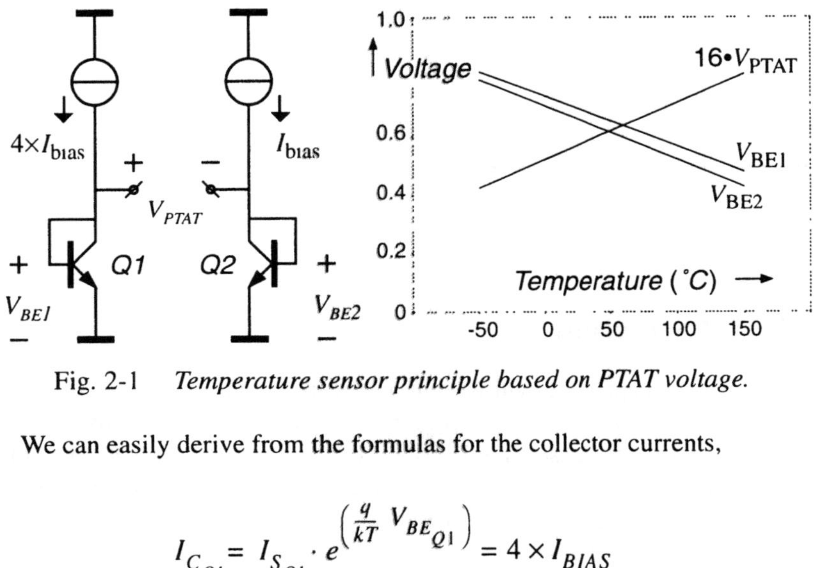

Fig. 2-1 *Temperature sensor principle based on PTAT voltage.*

We can easily derive from the formulas for the collector currents,

$$I_{C_{Q1}} = I_{S_{Q1}} \cdot e^{\left(\frac{q}{kT} V_{BE_{Q1}}\right)} = 4 \times I_{BIAS}$$
$$I_{C_{Q2}} = I_{S_{Q2}} \cdot e^{\left(\frac{q}{kT} V_{BE_{Q2}}\right)} = I_{BIAS} \qquad (2\text{-}1)$$

that a (PTAT) voltage proportional to the absolute temperature T results:

$$\Delta V_{BE} = V_{BE_{Q1}} - V_{BE_{Q2}} = \frac{kT}{q} \cdot \ln\left(4 \cdot \frac{I_{S_{Q2}}}{I_{S_{Q1}}}\right) \qquad (2\text{-}2)$$

In which k represents Boltzmann's constant, q the electron charge and I_S the transistor saturation current. The case of equation (2-2), using a current density ratio of four, and $I_{SQ1} \approx I_{SQ2}$, results in a small positive temperature coefficient of approximately 0.12 mV per Kelvin. The base-emitter voltage itself demonstrates a much larger but slightly less linear temperature dependency which can be expressed in the form of a Taylor expansion of the V_{BE} curve at a reference temperature T_R :

$$V_{BE} = V_{BG_{T=0}} + (\eta - m) \cdot \frac{kT_R}{q} + \lambda_{T=T_R} \cdot T - \frac{1}{2}(\eta - m) \cdot \frac{kT_R}{q} \cdot \left(\frac{T - T_R}{T_R}\right)^2 \tag{2-3}$$

In this equation V_{BG} represents the silicon bandgap voltage, λ the linear temperature coefficient which measures approximately -1.9 mV/K at T_R=300K, and (η-m) a physical parameter of about 3.5 which is somewhat process dependent. The voltage signals V_{BE} and V_{PTAT} are displayed in Fig. 2-1, where the PTAT voltage had to be magnified by a factor of 16 to obtain a comparable temperature coefficient. Both effects can be used for measuring temperature. In section 2.3.1 will be shown how to combine both signals in a fully integrated circuit and to circumvent the need for an external reference voltage.

2.2.2 Pressure sensor

A well-known principle for measuring pressure is by the use of a thin membrane or diaphragm, because a diaphragm tends to deflect under pressure. Micromachining techniques make it possible to fabricate thin mechanical structures on a silicon wafer, even without harming the integrated circuits processed on the same wafer [4,5,6,7,8,9]. A common technique, known as bulk-micromachining and also accomplished at DIMES, is to (KOH) etch the masked backside of the wafer (p-substrate) and obtain an electrochemical etch stop by applying a voltage to the n-epi or buried-n layer at the frontside. In that way the wafer thickness (~500µm) can be locally reduced, under an angle of approximately 55°, to epi-thickness (~3-7µm), whereas the horizontal dimensions can be controlled by the backside mask [10,11]. In this way, a structure as illustrated in the pictures of Fig. 2-2 can easily be fabricated.

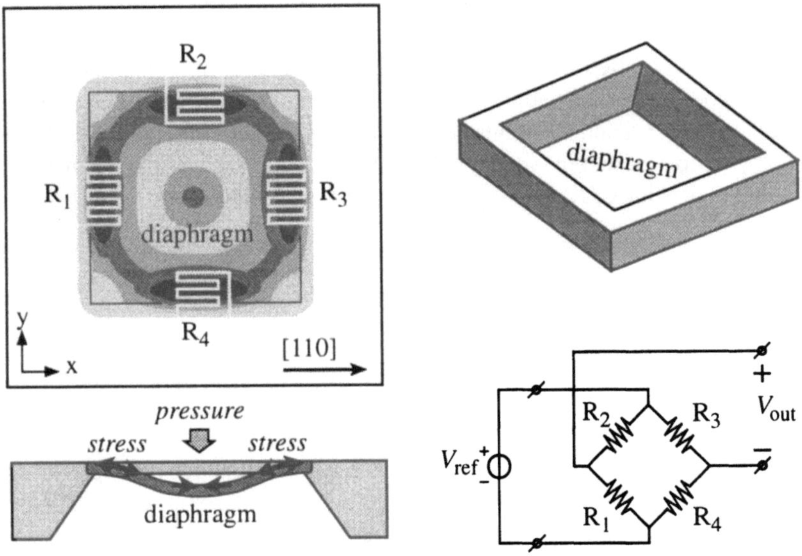

Fig. 2-2 *Pressure sensor principle with Wheatstone bridge.*

Now the art is in electrically sensing the pressure induced deflection of the micromachined diaphragm. A capacitive technique seems attractive but has the disadvantage that another capacitor plate is required, which has to be fabricated/mounted as close as possible to the diaphragm for maximum capacitance signal [12,13,14]. This is problematic considering the packaging and the electrical connections, as well as incompatible with some of the pressure sensor types indicated in Fig.2-3 because of the required backside opening [7]. Solutions making use of wafer-to-wafer or wafer-to-glass bonding are still under development [5,11,15,15].

An already well-developed method makes use of silicon piezo-resistors [7,16,17]. As indicated in the cross-section and top view of the sensor in Fig.2-2, the deflection of the diaphragm results in a shear stress at the surface of the diaphragm. Provided that the silicon lattice is correctly oriented, the piezoresistance effect of silicon can be used to sense the change in stress [4,17,18]. The topview also gives an impression of the (absolute) stress profile when under pressure; the darker areas have larger stress, and the stress-sensitive piezoresistors are thus placed in those areas. The orientation of the resistors is such that the stress results in an increase of resistance for R_1 and R_3 and a decrease for R_2 and R_4. The silicon resistors (shallow-*p* diffusion) can be fabricated using standard IC

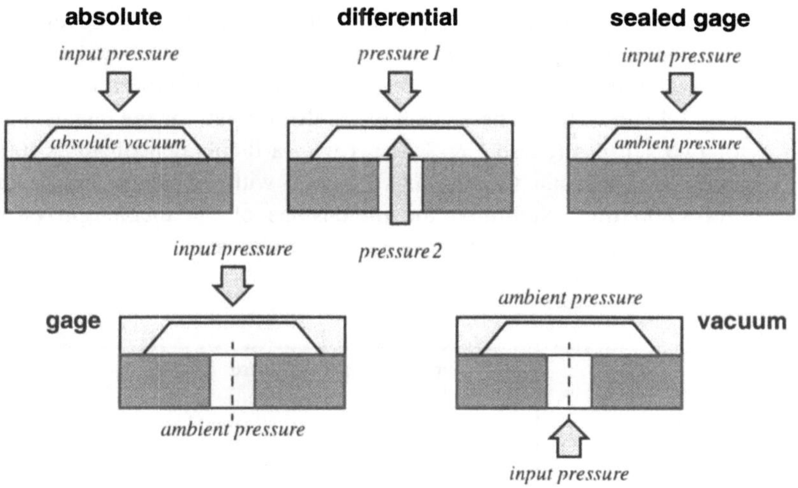

Fig. 2-3 *Different types of pressure sensors.*

processing (on the same wafer), and are connected in a Wheatstone bridge configuration, as shown in Fig.2-2. A pressure-related output voltage signal, V_{out}, proportional to the bridge supply voltage, V_{ref}, is thus generated [7].

2.2.3 Measurement errors

In the fabrication of sensors certain changes in process parameters and also in material (wafer) properties will have to be tolerated, and, ergo, a variation in the properties of the sensors, even when produced in the same process, must be expected. This usually results in a statistical (Gaussian) distribution of the errors in the transfer of the sensors with respect to the average sensor transfer. The manufacturer tries to get the average sensor transfer as close as possible to the desired sensor function, and to minimize the deviation of the transfer errors. Generally the spread in the error is much larger than the desired accuracy, and selection or calibration is required in order to obtain accurate sensors.

In case of the PTAT sensor the variations in the transfer are determined by the matching properties of standard electronic components. This will be examined in section 2.3.1.

In the framework of a master thesis' project on pressure sensor self-test some bulk-micromachined pressure sensors have been developed and realized in the DIMES-01 process. We will use those sensors as an example. Several of these devices have been measured using DRUCK test equipment. Fig. 2-4 shows the transfer, relative bridge output voltage versus pressure, of two sensors (Sensor1 & Sensor2) with a diaphragm of $400 \times 400\ \mu m^2$, and two sensors (Sensor3 & Sensor4) with a larger diaphragm of $800 \times 800\ \mu m^2$. Naturally, the dimensions of the diaphragm (size and thickness) determine the pressure sensitivity and the pressure range in which the sensors can be used.

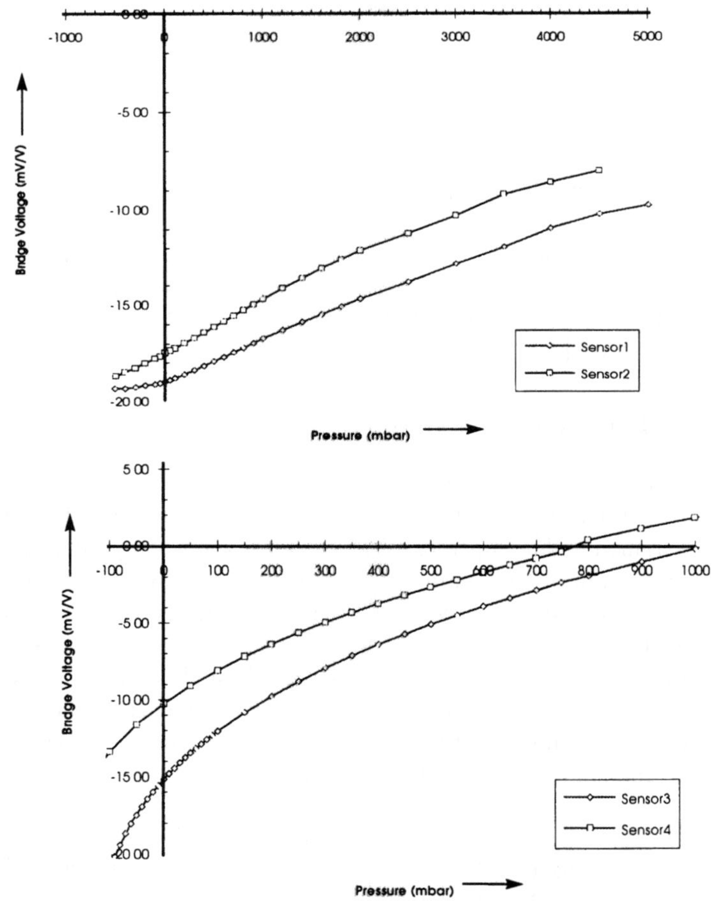

Fig. 2-4 *Measured transfer of pressure sensors developed in Dimes-01.*

Despite the fact that the sensors come from the same wafer, a significant difference in offset and sensitivity between two sensors is found, for both cases. An even larger difference can be expected when also taking into account the variation in temperature coefficients of offset and sensitivity. The large offset errors in the shown examples are mainly due to imperfections in the first resistor layout, which can easily be improved in a next version. Nevertheless, error sources such as initial stress, geometrical variations, doping concentration profiles, and resistor mismatch will result in a large error variations. The sensor construction does not allow to put resistors close together to improve matching. The source of the signal errors also explains why there is not a typical sensor-characteristic nonlinearity error but a large device-to-device variation. An overview of error contributions in piezoresistive pressure sensors can be found in [7:chapter 4] and [19]. A total error band of 30% of the full-scale can be a reasonable expectation.

Notwithstanding, one of the largest sensor markets is that for pressure transducers. Thus a strong push from that market is expected for stimulation of integrated calibration techniques. We have to conclude that in order to realize production of a pressure sensors with standard transfer from pressure range to output range, the calibration will have to compensate large error variations.

2.3 Analog Interface circuits

2.3.1 Temperature sensor readout

To enable the read-out of a temperature signal, as presented in section 2.2.1, by an AD-converter, another signal has to be generated which can be used as a reference. Hence, we will first describe the construction of a so-called bandgap reference. The temperature signal and the reference signal must be combined in such a way that a ratiometric signal is obtained which makes optimal use of the AD-converter's signal range. This will be explained next.

By summing a V_{BE} signal and a signal proportional to V_{PTAT} the first-order (linear) temperature coefficients can be cancelled, and a reference signal results, which is referred to as bandgap reference. As can be seen from equation (2-3) it relates directly to the bandgap voltage V_{BG}, although a small offset error and second-order error are present. Bandgap references and PTAT *current* sources are well-known circuits in IC design as they are often used for biasing purposes.

The left half of the circuit displayed in Fig. 2-5 shows a low-voltage PTAT source [20], which makes use of the ratio-ed transistors Q1 and Q2. These transistors are inserted in a control loop that establishes equal bias currents for both transistors. Because of the ratio in current density the difference between the base-emitter voltages is proportional to the absolute temperature, as already shown in equation (2-2). The value of the bias current is determined by the PTAT voltage at the emitter of Q2 and the value of resistor R_1. The current I_{VPTAT} inherits the proportional-to-absolute-temperature property of the voltage. The current mirror on the left-hand side only serves to generate a small start-up current to prevent a zero bias solution at start-up.

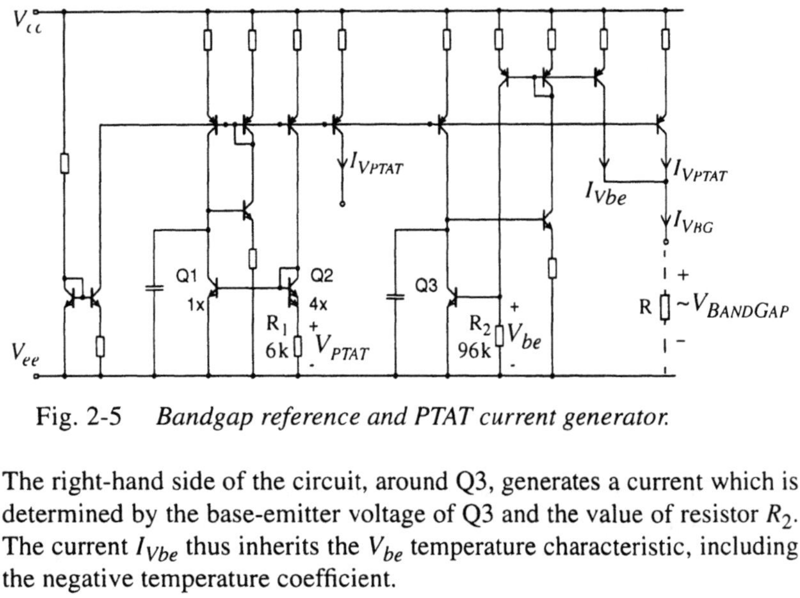

Fig. 2-5 *Bandgap reference and PTAT current generator.*

The right-hand side of the circuit, around Q3, generates a current which is determined by the base-emitter voltage of Q3 and the value of resistor R_2. The current I_{Vbe} thus inherits the V_{be} temperature characteristic, including the negative temperature coefficient.

The two current signals can easily be summed at the output, resulting in $I_{VBG}=I_{Vbe}+I_{PTAT}$. When inserting that current in a resistor R, it is converted back into a voltage signal. The voltage will then be proportional to the weighed sum of the PTAT voltage $V_{PTAT} \bullet R/R_1$ and the base-emitter voltage $V_{be} \bullet R/R_2$, which can be equalized to the silicon bandgap voltage.

A simulation of the circuit, see Fig.2-6, shows the PTAT signal with a positive temperature coefficient, the V_{BE} signal with a negative temperature coefficient, and the constant sum signal. By correctly dimensioning the ratio of the resistor values of R_1 and R_2, in this case a ratio of 16, the temperature coefficients of both current signals can be matched so that the sum of the currents, referred to as I_{VBG}, remains constant as a function of temperature.

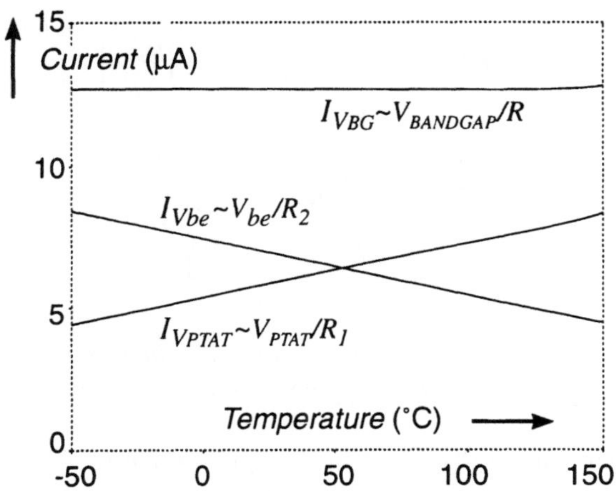

Fig. 2-6 *Current signals as function of temperature, using ideal resistors.*

The simulation shown in Fig.2-6, was done for ideal resistors, meaning that the resistance was considered independent of temperature. In reality the temperature coefficients of the resistors directly affects the curves of the *current* signals. However, when using the same kind of resistors for R, R_1, & R_2, the bandgap *voltage* signal will not be affected because it depends on the resistor ratios. Similarly, the effect of the temperature coefficients will cancel if we use the ratio of two current signals.

Temperature sensor circuit

For use in a smart temperature sensor with AD-converter we like to realize a circuit which generates two current signals of which the ratio varies approximately from 10% to 90% for the intended temperature range. Within the operating temperature range of IC technology, the ratio of the PTAT and the bandgap reference signal varies from 35% to 65%. Therefore, we prefer not to use that ratio directly, but to modify the current signals.

This can be done by combining the PTAT and bandgap currents, for example in the way shown in Fig.2-7. A fraction of the bandgap current is copied and subtracted from the PTAT current to generate a new temperature current signal I_T. A down-scaled copy of the bandgap current is used as a new reference current signal I_{REF}.

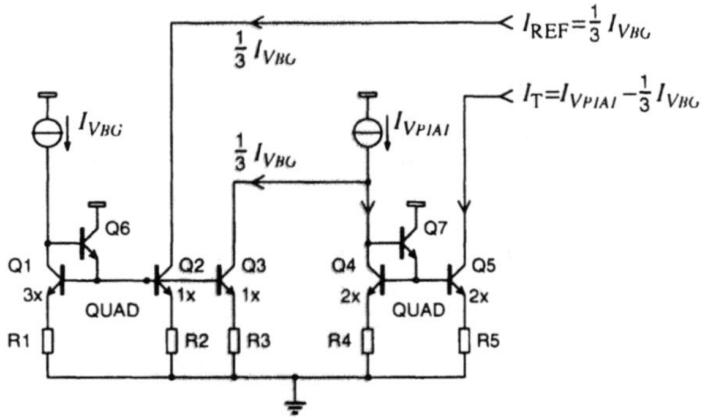

Fig. 2-7 *Temperature "sensor" circuit generating I_T and I_{REF}; I_{VBG} and I_{VPTAT} come from the circuit of* Fig.2-5.

The simulation displayed in the upper graph of Fig.2-8, shows the two currents as a function of temperature, now taking into account the temperature coefficients of the resistors in the IC process. It clearly shows

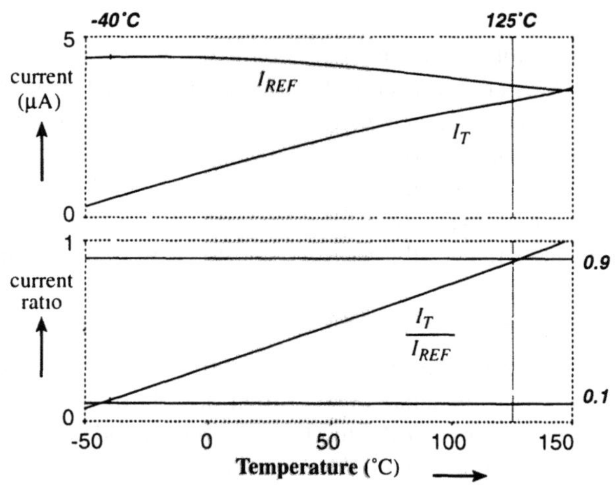

Fig. 2-8 *Ratio of I_T and I_{REF} used as measurement signal.*

that the absolute current I_{REF} is certainly not constant, and that I_T is not so linear. However, the ratiometric signal I_T/I_{REF}, displayed in the lower half of Fig.2-8, is good and is not affected by the temperature dependence of the resistors. The sensor circuit is dimensioned in such a way that the ratiometric signal range of [0.1-0.9] maps on the temperature range of -40 °C to +125 °C. The current mirrors in the sensor circuit of Fig.2-7 can be dimensioned differently to obtain a different temperature range, within the operation limits of ICs of course.

To study the effect of process variations on the signal, a simulation has been done for a worst-case relative mismatch of ±0.5% and an absolute variation of ±25%. Fig.2-9 shows the results on the ratiometric signal and the error with respect to the desired linear curve.

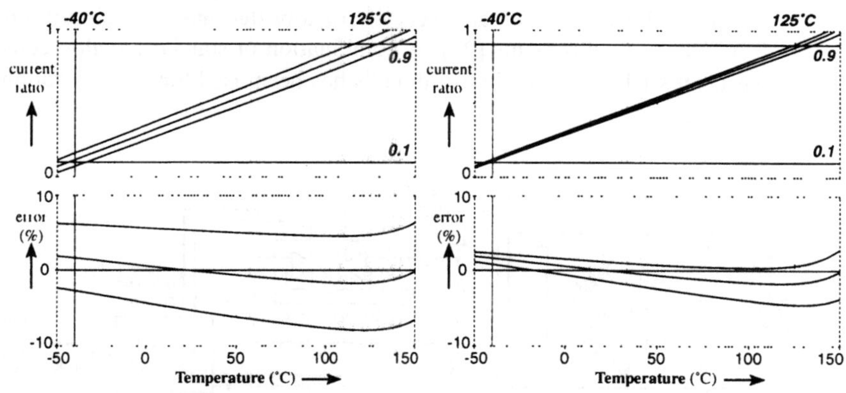

Fig. 2-9 *Effect on transfer and error curves resulting from worst-case relative (1%) and absolute (50%) variations.*

First of all, it demonstrates that the effect of relative errors is much larger. Second, it can be concluded that, in fabrication of such temperature sensors, a spread in offset and gain errors of several percents, say ±5%, should be expected. The linearity error will be relatively small, in the order of 1%, for the specified temperature range. To realize a silicon temperature sensor with exactly the desired transfer from temperature to ratiometric output, these errors will have to be calibrated. Furthermore, it has to be noticed that the nonlinearity curve has a shape that is characteristic to this type of sensor. The shape doesn't vary much from device-to-device. It is determined by the nonlinearity in the V_{BE} curve as expressed in equation (2-3). The temperature coefficient of the resistors is a second source for nonlinearity but results only in a very small device-to-device variation of the nonlinearity error.

2.3.2 Pressure sensor readout

The piezoresistive pressure sensors based on a Wheatstone bridge configuration inherently make use of a ratiometric conversion; as explained in section 2.2.2 a pressure signal will result in an output voltage signal related to the bridge reference voltage. The AD-conversion which follows the analog readout circuitry has to realize a conversion of this ratiometric signal, by using both the bridge output voltage and the bridge reference voltage. As seen in section 2.2.3, pressure sensors may display an offset which is in the same order of magnitude as the full-scale signal range. This offset can be eliminated by using a chopping technique. As an example, Fig. 2-10 gives an overview of a bridge interface circuit which fulfils these requirements [21]. A combination of standard analog cells of the DIMES-01 library and custom cells has been used for the realization of this circuit.

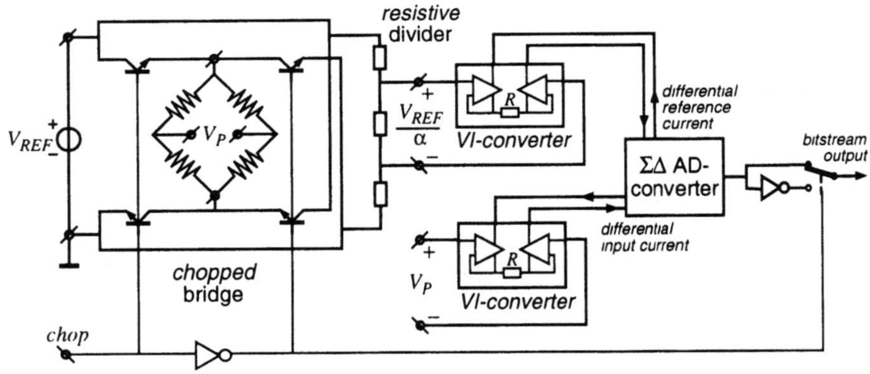

Fig. 2-10 *Pressure bridge readout circuit.*

The chopping is implemented by using saturated bipolar transistor switches which invert the polarity of the bridge reference voltage. Changing the polarity of the reference voltage, inverts the polarity of the bridge output voltage signal but not that of the bridge offset voltage. By subtracting the inverted and non-inverted signal the bridge offset is eliminated. This is preferably done at the end of the signal chain, as shown in the figure, so that all other offset contributions are removed as well.

As seen in section 2.2.3, the ratiometric bridge signal usually does not get much larger than 1-2%. To optimally use the dynamic range of the AD-converter it is necessary to scale down the bridge reference signal. In Fig.2-10 this is realized with, accurately scaled, and carefully laid-out resistors. Both the down-scaled reference voltage, and the bridge output voltage, are converted to currents, using VI-converters, and passed on to the sigma-delta AD-converter for a ratiometric conversion.

For the pressure sensor as well as for the temperature sensor we have shown how to obtain a current ratio proportional to the sensor signal. The same can be done for many types of sensors. The obtained current ratio can be accurately converted into a digital (bitstream) signal, using the continuous-time sigma-delta converter explained in the next section.

2.4 Analog-to-Digital conversion

Much effort has been, and is being put in comparing resolution, speed, and power consumption of several AD-conversion principles and architectures, however it does not fit in the scope of this work to discuss all the investigated AD-converter types. Information can be found in [22,23,24], for example. We will confine to summarizing the advantages of the preferred AD-converter type, the delta-sigma or sigma-delta converter[1]. Advantages of sigma-delta converters are:

- *serial one-bit digital clock-synchronous output*
- *simple and compact architecture which can be implemented using a small number of components*
- *a low-power consumption, especially at low sample speeds*
- *no accurate components or component matching required*
- *exchange of sample speed and resolution possible*
- *high resolution possible (16-bit)*

Disadvantageous can be the fact that the digital bitstream output needs digital post-processing in the form of an averaging, or noise-shape filter, in order to convert the one-bit output signal to multi-bit samples, which can

[1] Both names, delta-sigma and sigma-delta, are common, and refer to the same type of converter.

be used in a digital processor. Such a filter might be implemented in hardware on a separate (CMOS) chip which can be connected to the sensor bus, or in software on a microcontroller or PC bus master. This enables a shared use of the filter for all the smart sensors that produce a bitstream output, see also section 2.5.

We will first proceed by explaining the sigma-delta conversion principle.

2.4.1 Sigma-delta converter

The sigma-delta principle is based on oversampling, and on feedback and filtering (noise-shaping) of the quantization error. The operation principle in the time domain will be explained using Fig.2-11 and Fig.2-12.

The basic sigma-delta converter consists of an integrator, often implemented in a feedback configuration using an OpAmp, and a clocked comparator, mostly implemented as a master-slave flipflop.

The fact that we largely oversample the input signal,

$$f_{clock} \gg f_{Nyquist} = 2 \times bandwidth \tag{2-4}$$

permits the assumption that the input signal is considered constant for many clock periods.

The first figure illustrates the situation that the presumably constant input current is integrated on the integration capacitor, and results at the integrator output in a ramping voltage signal, with an upward slope proportional to the input signal. The clocked comparator compares this voltage signal to the comparator's reference level V_{cmp}, and outputs a digital decision, low or high, only once each clock period, for example at the positive clock edge. In this situation the output remains low until the first clock edge after the integrated signal exceeds the comparator's reference level. Then the comparator switches the output signal to high, and turns the switch of the reference current, as illustrated in the next figure. Provided that $I_{ref} > I_{in}$, an opposite integrator input current of $I_{ref} - I_{in}$ results now in a proportional negative slope at the integrator output. The comparator output is high until the integration signal has dropped below the reference level, and a clock edge occurs. The output will become low again, and the sigma-delta converter switches back to the situation of Fig.2-11, and the procedure repeats itself.

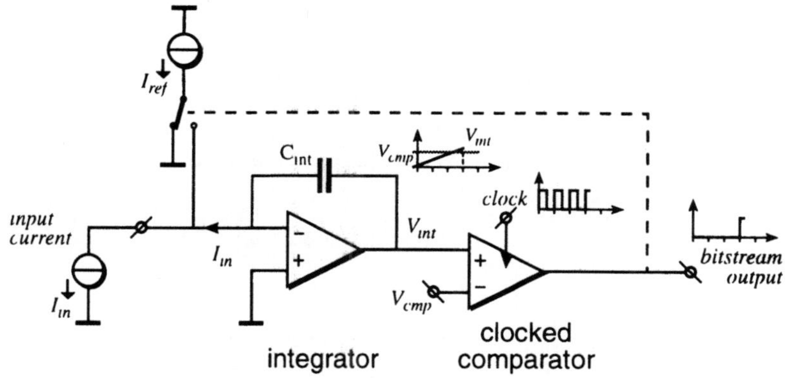

Fig. 2-11 *Sigma-delta converter principle, output low (A).*

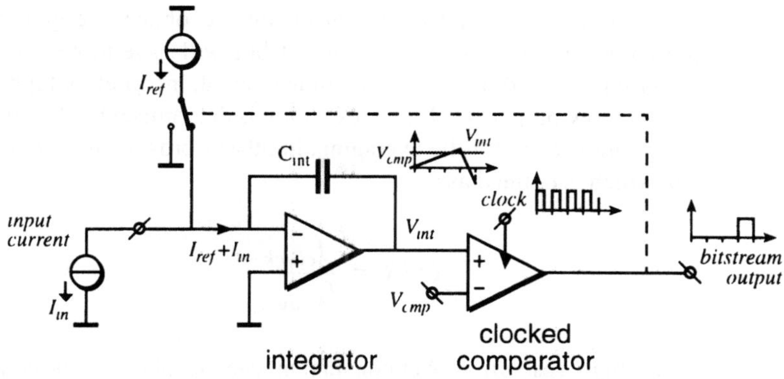

Fig. 2-12 *Sigma-delta converter principle, output high (A).*

When considering a larger number of clock periods we see that the integrator signal oscillates around the reference level, and the digital output produces a series, or pattern of ones and zeroes at the beat of the clock signal, as depicted in Fig.2-13 for an input signal of $I_{in}=0.19 \cdot I_{ref}$.

The input signal range is usually limited to the signal range from $0.1 \cdot I_{ref}$ to $0.9 \cdot I_{ref}$, to avoid very low slopes in the integrator signal. Through simple reasoning it can be found that a large input signal results in steeper upward slopes, and less steep downward slopes, and thus a large number of high outputs. The reverse applies for small input signals; the number of high outputs will be small. Hence the output pulse density seems proportional to the input signal; in fact the value of the input signal

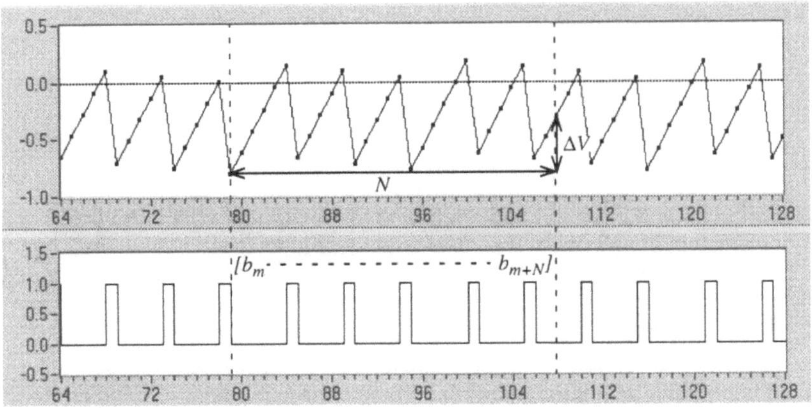

Fig. 2-13 *Sigma-delta integrator and bitstream signal.*

is encoded in the digital output bitstream. To obtain a digital number representing the input signal, the output bits will have to be counted (or otherwise processed) over a certain time interval, interpreting high outputs as *+1* and low outputs as *0*. The resolution is determined by the number of clock periods taken for the averaging, in other words by the *oversampling ratio*, which is defined as:

$$OSR = \frac{f_{clock}}{f_{Nyquist}} \tag{2-5}$$

A very high resolution AD-conversion can be obtained provided that integrator "leakage", clock jitter, delay, and hysteresis are minimized [24,...,29]. To fulfill those requirements accurate components are not required, as opposed to the requirements for other AD-converter types. Besides the *OSR* and the noise-shape order, the performance is determined by the circuit design and the invested bias power.

By realizing the sigma-delta converter in a fully differential topology, as depicted in Fig. 2-14, common-mode and supply interference, important error sources in instrumentation can be strongly suppressed. This configuration allows a differential input signal limited in signal range from $-0.9{\cdot}I_{ref}$ to $+0.9{\cdot}I_{ref}$. High outputs should now be interpreted as *+1*, and low outputs as *-1*. As often seen in practical realizations, the two displayed reference sources can be replaced by a single reference current of $2{\cdot}I_{ref}$, provided that a common-mode (CM) control loop at the input of the

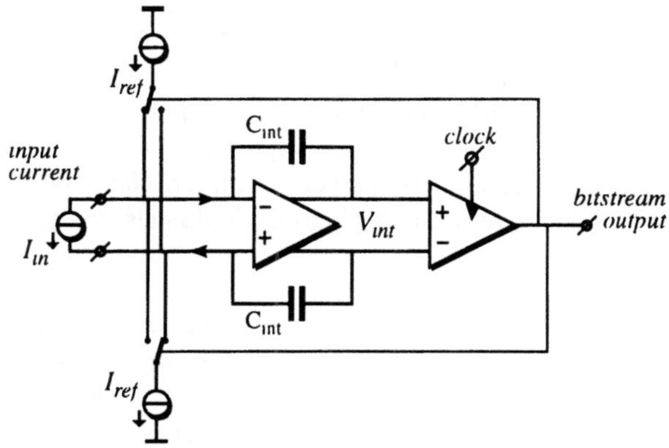

Fig. 2-14 *Differential sigma-delta converter.*

integrator takes care of fixing the input CM-level. This is necessary to obtain a *symmetrical* differential integration signal. The clocked comparator in this configuration simply decides the *polarity* of the differential signal, it compares the differential signal with the intrinsic reference of *0*. Fig.2-15 further demonstrates the operation of the (differential) sigma-delta converter. A slowly varying sine-wave input signal clearly results in a pulse-density modulated bitstream signal. In this example it is attempted to reconstruct the original signal by taking the average of the bitstream signal in clusters of 16 clock periods.

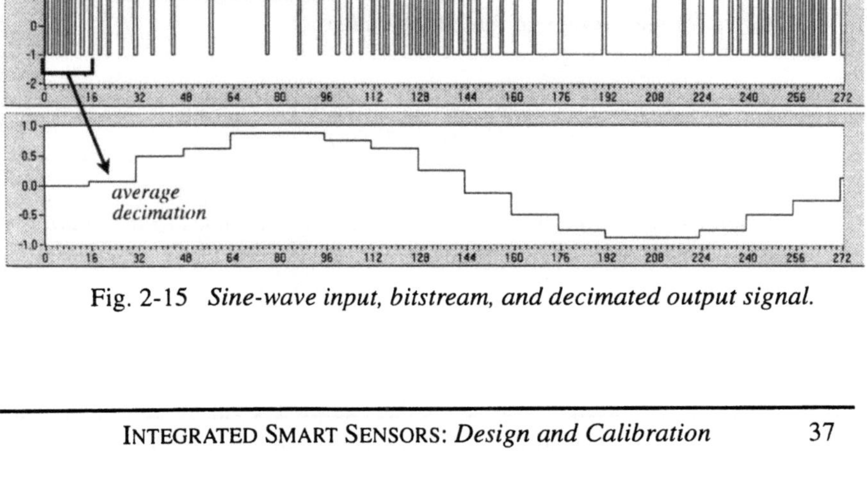

Fig. 2-15 *Sine-wave input, bitstream, and decimated output signal.*

Essentially the principle of sigma-delta conversion is based on simple (one-bit) quantization of the (filtered) input signal, whereas the quantization error is fed back to the input, and averaged (filtered) to zero (for DC) by signal integration. The output signal is composed of the digitized input signal at low frequencies, and the quantization (noise-shaped) error at high frequencies. A more elaborate explanation of the sigma-delta converter, using frequency domain analysis, will be given in *chapter 5.*

Another approach is to look at the *charge balance* which is obtained: on the long term the signal at the output of the integrator averages to zero (i.e. the remaining charge error can be neglected). This means that over a long interval the integral of the sum of all integrator input signals is zero, or in the discrete time domain:

$$\sum_{n=m+1}^{m+N} I_{in}[n] - \sum_{n=m+1}^{m+N} b_n \cdot I_{ref} = \frac{C \cdot \Delta V}{T_{clk}} \approx 0 \qquad (2\text{-}6)$$

in which N is the number of clock periods in the interval, and b_n represent the corresponding series of output bits, referring back to Fig.2-13. The remaining charge, determined by the difference ΔV between the integrator voltage, at the start and at the end of the interval, and the integrator capacitor C, can be neglected if the time interval $N \times T_{clk}$ is large with respect to the clock period T_{clk} : $N >> 1$. The maximum voltage error results when the minimum/maximum input signal is just not detected by the comparator, thus $\Delta V_{max} = I_{ref} \times T_{clk} / C$. Which means the maximum error on the right-hand side of equation (2-6) corresponds to $1 \times I_{ref}$, whereas the maximum on the left-hand side of the equation is N times larger.

Provided the input signal remains constant during the interval, we can derive for the average of the bitstream output:

$$\frac{1}{N} \sum_{n=m+1}^{m+N} b_n = \frac{I_{in}}{I_{ref}} \qquad (2\text{-}7)$$

which confirms that the output bitstream signal relates to the *ratio* of the input current and the reference current.

2.5 Digital bus or microcontroller interface

2.5.1 Microcontroller compatible interfaces

The bitstream output of a smart sensor with a sigma-delta converter is microcontroller compatible in the sense that it can easily be read in, and interpreted by a microcontroller or PC, as it is a serial, one-bit, digital, and clock-synchronous signal [30]. The appropriate sigma-delta clock signal can be generated by the microcontroller, and is sent to the sigma-delta. The digital bitstream signal is sent from the sigma-delta to the microcontroller. The microcontroller processes the bitstream in order to obtain multi-bit samples with the desired resolution. In order to read out multiple smart sensors with bitstream outputs the microcontroller needs to be equipped with a multiplexer which enables the selection of the different sensors; the clock line might be shared by several smart sensor devices. This situation is pictured on the left side of Fig. 2-16.

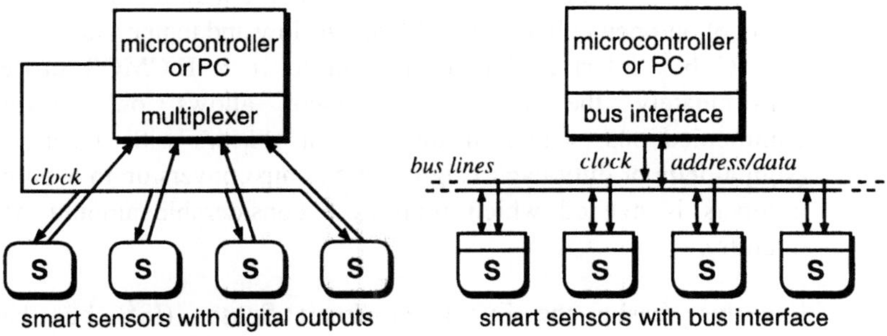

Fig. 2-16 *STAR-coupled versus BUS architecture.*

It is clear that the with the number of sensors also the wiring complexity increases. Other disadvantages of this *star-coupled* architecture are that the number of sensors is limited by the multiplexing capabilities of the microcontroller (usually 8 to 16), that the position of the sensor determines the multiplex address, and thus sensors cannot be swapped, and that only direct sensor-to-master communication is possible.

The *bus* architecture shown on the right-hand side of Fig. 2-16, overcomes all these disadvantages. The sensors and the microcontroller/PC are equipped with a bus interface, and connected by common lines, a *clock line* and a *data line* for example. A *bus-slave* sensor will put its data on the bus data line only after it has been correctly *addressed* by the microcontroller/PC *bus master* according to a *bus protocol*. The sensor data can be read by the master but also by other devices on the bus, a digital bitstream filter for example. Bidirectional communication is also possible, for example to program the calibration factors of a sensor before reading it out. Other advanced features which become available in a bus architecture are sensor identification and alarm interrupts.

2.5.2 Smart Sensor bus interfaces

Because it often is not possible to integrate complex digital circuitry with the sensor device, the first requirements for a smart sensor bus is minimum complexity at the bus-slave side. Another requirement is to have a minimum number of wires, because for use in sensor systems each connection introduces extra fabrication costs. A sensor bus which fulfills these requirements is the Inter-Integrated Circuit (I^2C) bus, which has been developed by Philips as internal bus for example for consumer products with microprocessor control. The bus has two supply wires and two wires for communication, using a simple addressing and handshake protocol on open collector (OC) lines. A flow and temperature sensor with an I^2C bus interface has been realized in a BiCMOS process [31]. Unfortunately the I^2C bus protocol allows only serial byte communications, which means that for sensors with a serial one-bit (sigma-delta or duty-cycle) signal an on-chip conversion to 8-bit or 16-bit numbers is needed which requires a considerable amount of digital circuitry.

For this reason the Integrated Smart Sensor (IS^2) bus has been invented [32]. The IS^2 bus is very similar to the I^2C: it also has two supply lines and two serial communication lines, but it has an open data transmission format [33,34,35]. In addition to the open collector technique Manchester encoding is used, so that in combination with the clock four different data states can be distinguished, as illustrated in Fig.2-17.

The clock and the data line are both pulled up by resistors, which are connected to the positive supply line. Using a strong open collector driver (R_{clk}≈500Ω), the master puts a highly stable clock signal on the clock line, which is indispensable for the accuracy of the sensors' (time-continuous) sigma-delta converters. The data line can be pulled down by any device on the bus, using a less powerful open collector

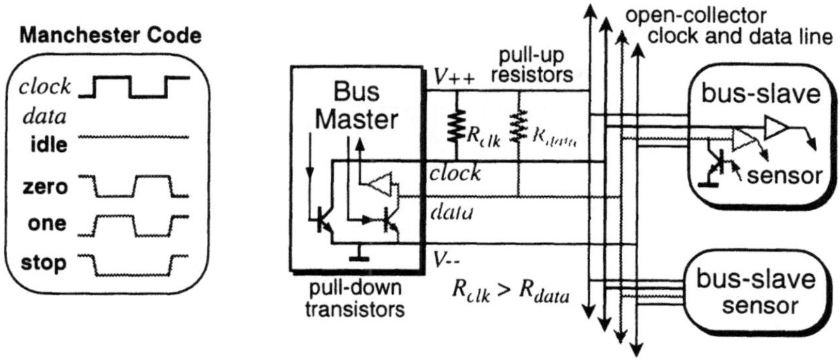

Fig. 2-17 *Using a Manchester encoding in combination with open-collector lines.*

transistor (R_{data}≈5kΩ). Both the clock and the data line are read by the bus devices using high-ohmic input buffers.

When not a single bus device is active the data line remains high, and the bus is in idle state. After at least two idle states, a bus master can address a (sensor) bus slave by serially transmitting a start bit, eight address bits, and awaiting an acknowledge bit from the addressed bus slave, as explained in the bus protocol shown in Fig. 2-18. The slave can then put its data on the data line.

Digital clock-synchronous data is transmitted according to the Manchester code described in Fig. 2-17, a *zero* means pulling down the data line during the high phase of the clock, and a *one* means pulling down the data line during the low phase of the clock. When a *one* and a *zero* collide, because two sensors accidentally have responded to the same address, the line remains low during the whole clock period which corresponds to a *stop* signal which will be detected by the active bus devices, and reset the bus communication. This is also the way the master or an alarm sensor can send an interrupt. The length of a digital transmission is controlled by the master, so that it is possible to communicate bytes as well as any arbitrary number of bits, for example a long series of sigma-delta output pulses.

As mentioned before, it is also possible to put other kind of data on the bus. In that case the master keeps the clock line still (high) to prevent cross-talk distortion, and the slave can put an asynchronous digital signal (pulse-width or pulse-frequency modulated) or an analog voltage on the data line. It is up to the master to correctly interpret the data. Interruption

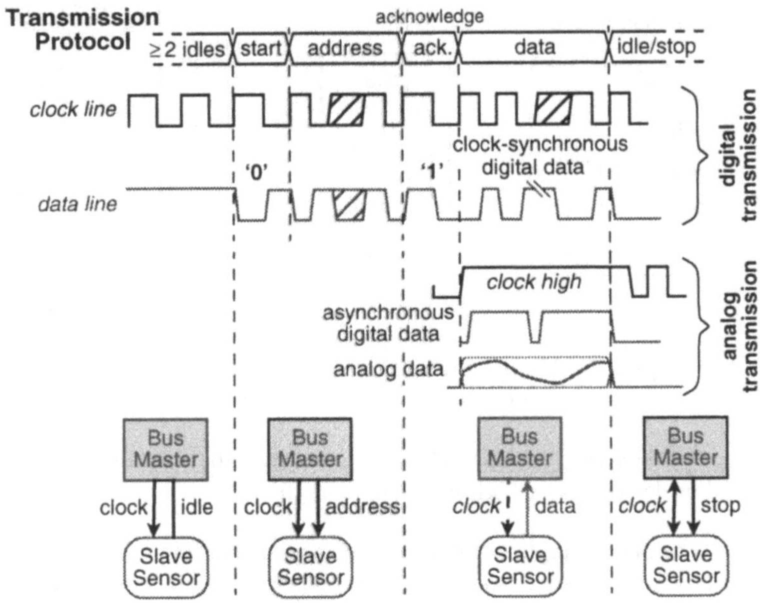

Fig. 2-18 *Explanation of the IS² bus communication protocol.*

is now only possible by (shortly) pulling the clock line down which should "wake up" the master, and get the clock running again.

It is also possible to make bus slaves receive data, for example to program a sensor's input range. At the side of the master the bus addresses will have to be organized in such a way that the master knows what kind of data to expect from or to send to the devices on the bus. This could be done by setting up a table with information about the sensors attached to the bus, containing address, type of data, number of samples, etc. or by splitting up the address range in sub-ranges for analog sensors, for sensors with bitstream outputs, etc. With eight address bits the maximum number of addresses is 256.

The hardware at the sensor (bus-slave) side is very simple. The scheme is shown in Fig.2-19. A Manchester decoder is needed to detect the status of the data line. After a start condition has been detected the decoded data bits are shifted in a register. A digital comparator is needed to compare the received address to the slave's own address. After positive comparison the slave immediately returns an acknowledge bit, and than puts its data on the bus. For digital data a Manchester encoder and OC driver are needed, and for other data an analog switch and a voltage driver. The slave set-up is thus very simple, and the necessary digital circuitry is so limited (approximately 20 gates and 12 flipflops) that it can be integrated in CMOS on a very small surface, but also in bipolar technology using for example Emitter-Coupled Logic (ECL) as will be demonstrated in *chapter 4.*

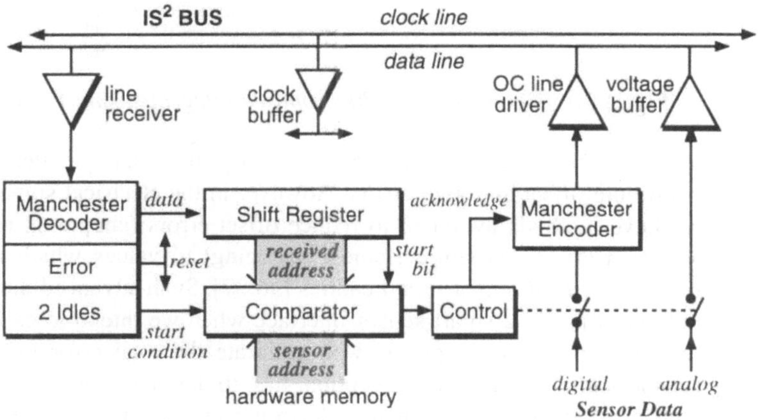

Fig. 2-19 *Configuration of an IS^2 bus-interface, for a bus slave.*

2.6 Integrating a calibration function in the smart sensor concept

The favored smart sensor configuration, for bipolar implementations at least, is reviewed in Fig. 2-20. Besides the (silicon) sensor it consists of an analog interface which realizes a sensor signal current and a reference current that are converted by the sigma delta converter into a digital bitstream signal. Next, the bitstream signal can easily be passed on by the IS^2 bus interface to the system's microcontroller or another bus master for example a PC. For many sensor types this scheme is probably the simplest way to realize a high-resolution AD-conversion and BUS interface

2.6.1 Error sources in the smart sensor signal path

Besides errors in the transfer of the sensor, from the physical input to the electrical output signal, we can also expect errors in the electrical transfer

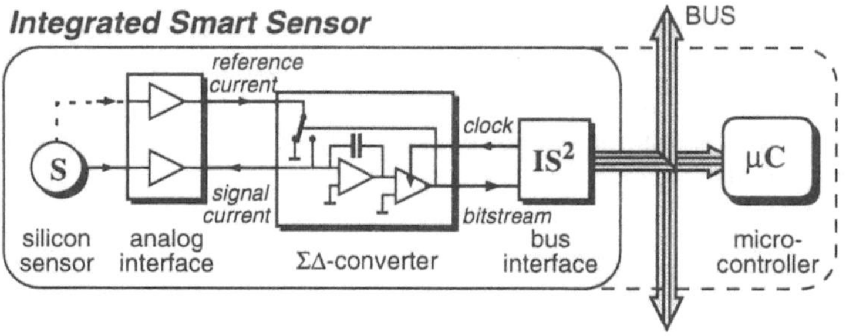

Fig. 2-20 *Signal chain of the proposed integrated smart sensor.*

of the other function blocks. Especially the analog interface may contribute offset and gain errors. However in the electrical signal domain we have methods available to reduce offset errors (chopping, auto-zero) and gain errors (dynamic element matching) to values which are much better than most sensor specifications [36,37]. Such advanced methods are not necessary in a smart sensor interface when we intend to calibrate the sensor anyway: we might as well calibrate the total error in the smart sensor signal chain. An exception has to be made when we expect considerable drift errors in the electronic circuit: in that case dynamic chopping or element matching methods will be useful.

2.6.2 Conventional sensor calibration techniques

Despite the smart sensor concept, most sensors are still fabricated (and distributed) separately from the electronic interface. Therefore, conventional calibration techniques concentrated on correction of the sensor transfer itself, during or just after fabrication. An example is the use of laser-trimmed resistors in a pressure sensor bridge, see section 3.2.1. A disadvantage of this approach is that it is difficult to automate the calibration, and that it is thus an expensive procedure in a fabrication line. Furthermore, it means that the electronic interface must match the specifications of the calibrated sensor, which could require a separate calibration or a more advanced circuit technique.

On the system level, sensor calibrations are usually performed on the systems computer. Data-acquisition programs, such as LabView, offer many possibilities for an advanced digital calibration. Disadvantage of this approach is that each new or replaced sensor in the system first has to be calibrated by the user, rather than by the manufacturer, which means the user needs (expensive) reference equipment. Furthermore, the calibration requires computational power of the system, and the calibration information is not attached to the sensor but stored in the system's computer. Altogether it will not stimulate the use of sensors in consumer products if such a calibration technique is needed.

2.6.3 Correction of the sensor signal transfer

As the objective is to find a way to integrate a calibration function in the smart sensor concept, we will start by looking for different possibilities to change the signal transfer of one of the different function blocks that have been presented. In the smart sensor signal chain shown in Fig. 2-20 we can choose to alter the transfer in:

- *the analog signal domain*
- *the AD-converter*
- *the digital signal domain*

In all cases it is important that the calibration is made digitally programmable. This will make it possible to fully automate the calibration in a fabrication process. It is also important that the errors of the whole signal chain are calibrated, even when we apply the correction at the front of the chain. This is done by taking the final (digital) output signal when measuring the sensor transfer during the calibration phase.

The three options will be further elaborated in the succeeding chapters. The sensor signal can be passed through an analog circuit with programmable offset, gain, etc. To realize a digital programmability for analog signals Digital-to-Analog Converters (DACs) are needed. This will become clear in *chapter 4*. In the sigma-delta AD-converter we have a mixed signal circuit which offers different possibilities for a programmable correction as will be clarified in *chapter 5*. Concerning the digital signal domain, it is often not possible to integrate advanced digital circuitry on the sensor chip. Alternatively it is possible to hybridly integrate a microcontroller die in the sensor package, or to use the microcontroller bus master for a digital calibration. Therefore, *chapter 6* concentrates on the realization of an advanced calibration function in the form of microcontroller software.

Before looking at those different practical implementation forms, the important aspects of calibration, and different linearization methods will be explained in *chapter 3*. Also an attractive polynomial calibration technique will be proposed.

2.6.4 Calibration memory

An issue that has not yet been addressed is the storage of the correction factors after calibration of the sensor. Obviously it is desirable that the correction is stored in a non-volatile memory integrated with the sensor or interface chip. Possible forms of nonvolatile memories, such as (E)PROM, zener or metal fuses, are mostly not available in standard IC processes. The development of such a memory in the used process could easily turn into a separate research project, and has not been given any attention within this project. The work in this book concentrates on the development of the circuitry needed to execute the correction. For the storage of the calibration coefficients simple volatile RAM is used, in the form of digital flipflop cells. The advantage of developing *digital* calibration circuitry is that the memory can easily be replaced with whatever digital memory is or becomes available in the used IC process.

References

[1] J.H. Huijsing, F.R. Riedijk, and G. v.d. Horn, "Developments in integrated smart sensors", *Sensors and Actuators A*, Vol.43, 1994, pp.276-288.

[2] G.C.M. Meijer, "Thermal sensors based on transistors", *Sensors and Actuators,* Vol.10, 1986, pp.103-125.

[3] G.C.M. Meijer and A.W. van Herwaarden, *Thermal Sensors*, Institute of Physics Publishing, Bristol/Philadelphia, 1994.

[4] S. Middelhoek and S.A. Audet, *Silicon Sensors*, Academic Press, London, 1989.

[5] R.F. Wolffenbuttel, "Fabrication compatibility of integrated silicon smart physical sensors, *Sensors and Actuators A*, Vol.41-42, 1994, pp.11-28.

[6] K.D. Wise and N. Najafi, "The coming opportunities in microsensor systems", *Proceedings Transducers 1991*, San Fransisco, USA, pp.1-7.

[7] J. Bryzek, K. Petersen, J.R. Mallon, L. Christel, F. Pourahmadi, *Silicon Sensors and Microstructures*, NovaSensor, Fremont, USA, 1990.

[8] J.H. Huijsing, "Signal conditioning on the sensor chip", *Sensors and Actuators*, Vol.10, 1986, pp.219-237.

[9] R. Frank, "Pressure sensors merge micromachining and microelectronics", *Sensors and Actuators A*, Vol.28, 1991, pp.93-103.

[10] L.K. Nanver, E.J.G. Goudena, and H.W. van Zeijl, "Dimes-01, A base-line bifet process for smart sensor experiments", *Sensors and Actuators A*, Vol.36, 1993, pp.139-147.

[11] R.P. van Kampen, *Bulk-Micromachined Capacitive Servo-Accelerometer*, PhD Thesis, Delft University Press, Delft, 1995.

[12] F. Schnatz et al., "Smart CMOS capacitive pressure transducer with on-chip calibration capability", *Sensors and Actuators A*, Vol.34, 1992, pp.77-83.

[13] C. Sander, J.W. Knutti, and J. Meindl, "A monolithic capacitive pressure transducer with pulse-period output", *IEEE Trans. Electron Devices*, Vol. ED-29, 1980, pp.42-48.

[14] L.S. Lee and K.D. Wise, "A batch-fabricated silicon capacitive pressure transducer with low temperature sensitivity", *IEEE Trans. Electron Devices*, Vol. ED-29, 1982, pp.42-48.

[15] T. Rogers and J. Kowal, "Selection of glass, anodic bonding conditions and material compatibility for silicon-glass capacitive sensors", *Sensors and Actuators A*, Vol.46-47, 1995, pp.113-120.

[16] S. Ansermet, D. Otter, R.W. Craddock and J.L. Dancester, "Cooperative development of a piezoresistive pressure sensor with integrated signal conditioning for automotive and industrial applications", *Sensors and Actuators A*, Vol.21-23, 1990, pp.79-83.

[17] D. Tandeske, *Pressure Sensors: selection and application*, Marcel Dekker Inc., New York, 1991.

[18] Y. Kanda, "piezoresistance effect of silicon", *Sensors and Actuators A*, Vol.28, 1991, pp.83-91.

[19] S. Kim and K.D. Wise, "Temperature sensitivity in silicon piezoresistive pressure transducers", *IEEE Trans. Electron Devices*, Vol. ED-38, 1983, pp.802-810.

[20] M. Gunawan, G.C.M. Meijer, J. Fonderie, and J.H. Huijsing, "A curvature-corrected low-voltage bandgap reference", *IEEE Journal of Solid-State Circuits*, Vol.28, No.6, 1993, pp.667-670.

[21] L.G. Fasoli, F.R. Riedijk, and J.H. Huijsing, "A general circuit for resistive bridge sensors with bus compatible output", *Proceedings IMTC'96*, June 1996, pp.1205-1210.

[22] R. v.d. Plassche, *Integrated Analog-to-Digital and Digital-to-Analog Converters*, Kluwer Academic Publishers, Boston/Dordrecht/London, 1994.

[23] P.R. Gray, B.A. Wooley, and R.W. Brodersen, edits., *Analog MOS Integrated Circuits II*, IEEE press, 1988.

[24] J.C. Candy and G.C. Temes, *Oversampled Delta-Sigma Data Converters*, IEEE Press, New York, 1992.

[25] F.R. Riedijk, *Integrated Smart Sensors with Digital Bus Interface*, PhD Thesis, Delft University Press, Delft, 1993.

[26] B.E. Boser, *Design and implementation of oversampled analog-to-digital converters*, PhD Thesis, Stanford University, 1989.

[27] P.J.A. Naus and E.C. Dijkmans, "Low signal-level distortion in sigma-delta modulators", *Proc. Audio Eng. Soc.*, March 1988.

[28] O. Feely and L.O. Chua, "The effect of integrator leak in sigma-delta Modulation", *IEEE Trans. Circuits and Systems*, Vol.38, No.11, November 1991, pp.1293-1306.

[29] R.W. Adams, "Design and implementation of an audio 18-bit analog-to-digital converter using oversampling techniques", *J. Audio Eng. Soc.*, Vol.34, No.3, March 1986, pp.153-166.

[30] A.W. van Herwaarden and R.F. Wolffenbuttel, "Introduction to sensors compatible with microprocessors", *Microprocessors and Microsystems*, Vol.14, No.2, 1990, pp.72-82.

[31] E.J. Hogenbirk, H.J. Verhoeven, F.R. Riedijk, and J.H. Huijsing, "an integrated smart sensor for flow and temperature with i^2c bus interface based on thermal sigma-delta modulation", *Proceedings Transducers 1993*, Yokohama, Japan, pp.792-795.

[32] J.H. Huijsing, R.F. Tuk, F.R. Riedijk, M. Bredius, and G. v.d. Horn, Mixed Analog/Digital Two-Line Bus System, Dutch Patent Application, Nr.93201595-1, June 1993.

[33] F.R. Riedijk R.F. Tuk, M. Bredius, and J.H. Huijsing, "A bus-controlled interface environment for an accurate read-out and flexible interconnection of integrated smart sensors in local systems", *Proceedings Transducers 1993*, Yokohama, Japan, pp.327-330.

[34] M. Bredius, F.R. Riedijk, G. v.d. Horn, and J.H. Huijsing, "The Integrated Smart Sensor (IS^2) bus", *Proceedings Sensor Expo West*, Philadelphia, USA, 1993, pp.243-247..

[35] F.R. Riedijk and J.H. Huijsing, "Sensor interface environment based on sigma-delta conversion and serial bus interface", *SENSORS, journal of applied sensing technology, Sensors Expo issue,* April 1996.

[36] F.M.L. van der Goes and G.C.M. Meijer, "A novel low-cost and accurate multi-purpose sensor interface with continuous auto-calibration", *Proceedings IMTC'96*, June 1996, pp.782-786.

[37] P.C. de Jong and G.C.M. Meijer, "Absolute voltage amplification using dynamic feedback control", *Proceedings IMTC'96*, June 1996, pp.67-71.

Calibration and Linearization Techniques 3

3.1 Introduction

In the first part of this chapter we will identify and distinguish the different types of errors which affect the transfer of the sensor, and explain which errors can be corrected by calibration. In the following section we will explain different linearization techniques which can be used to calibrate the offset, gain, and linearity errors in the sensor transfer. In the last part we will propose and explain a polynomial calibration method which can be used to calibrate and linearize the sensor transfer in a step-by-step approach. It will be shown how the method can be expanded to a two-dimensional polynomial calibration to be used for calibration of a cross-sensitivity error.

3.1.1 Errors in the sensor transfer curve

In sensor production it is desired to produce large batches of sensors at a time, and to have all of the sensors respond according to the same well-defined transfer within a certain accuracy. To verify this the sensors could be measured individually or by sample. When measuring the transfer of a sensor we may be confronted with one or more of the following error types [1,2]:

- ***offset**: when a zero (or minimum) physical input signal is applied, the measured output signal is not zero (or minimum of output range) but shows an 'offset' value.*
- ***gain, range, or full-scale error**: the sensitivity of the sensor is not as intended, the maximum physical input signal does not match with the maximum electrical output signal.*
- ***nonlinearity:** the sensor output does not change linearly with the physical input signal.*
- ***cross-sensitivity**: the transfer curve changes when measured at different ambient conditions (temperature, for example), thus the sensor is not only sensitive to the input signal but also to other parameters.*
- ***hysteresis**: the sensor transfer curve is different for decreasing physical signals than for increasing signals, once the signal passes a certain level.*
- ***drift**: the sensor transfer curve changes (slowly) in time.*

The errors are explained graphically in Fig.3-1. Generally, some of these errors are larger than the desired accuracy allows. To correct these errors, all sensors should be measured and calibrated individually to guarantee the accuracy.

Gain, offset, and also linearity and cross-sensitivity errors may be observed for practically all types of sensors. Hysteresis and drift errors need to be examined for some types of sensors, but can be neglected for other sensors. Although it may be somewhat complicated, hysteresis errors could be calibrated. However, other sensor-specific solutions should be considered first as they will often be less complicated or provide better results. Drift errors are difficult, if not impossible, to compensate or calibrate as it would require *prediction* of the future errors (extrapolation), and *time* measurement. Sometimes it is possible to reduce drift of the sensor by means of self-calibration, using an additional transducer (sensor or actuator) with a lower drift, see section 1.4.4, Fig.1-8, and Fig.1-9. For many sensors it is acknowledged that in a certain time (~*n* years) the sensor transfer curve may drift out of the specified range of accuracy, and *re*calibrations will be necessary. The smaller the drift errors, the longer the time span that can be allowed between recalibrations.

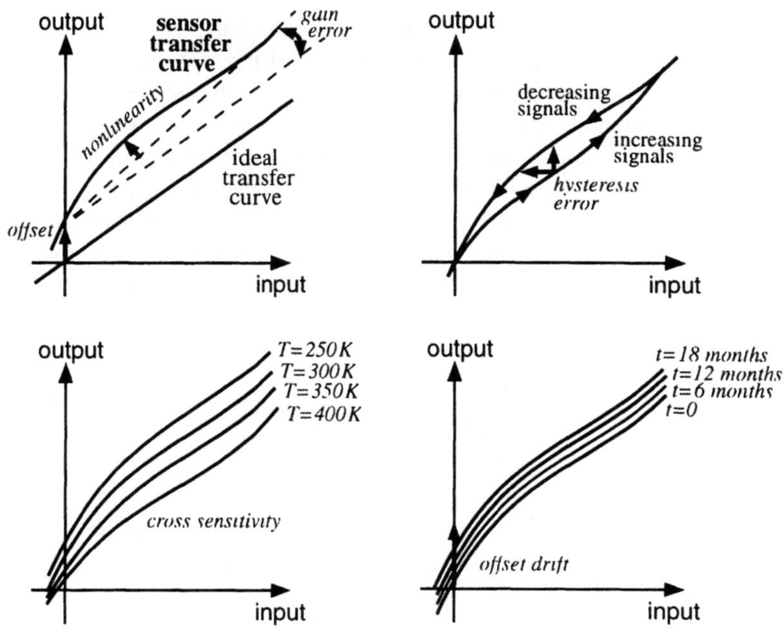

Fig. 3-1 *Possible errors in sensor transfer curve.*

The treatise presented in this book is limited to calibration of the offset, gain, linearity, and cross-sensitivity errors.

3.1.2 Systematic and random errors

Some errors may be characteristic for the type of sensor that is considered. It can be described in a model of the sensor. A thermal flow sensor, for example, shows a typical nonlinear rather than a linear transfer from flow velocity to sensor output signal. The source of such systematic errors should be traced in the sensor construction or read-out electronics. Other errors show a statistical distribution, due to random variations in the production process. For example, the offset of a resistive sensor bridge might show a Gaussian distribution because of uncontrolled and unpredictable differences in the values of the four resistors in the bridge.

When the errors in the transfer are larger than the desired sensor accuracy, the sensors in production need to be calibrated (or selected on measured specifications). Both systematic errors and random errors could be calibrated for each individual sensor. However, when the systematic error is much larger than the deviation of the random errors, this approach is not optimal. It is better to trace the source of the systematic error and correct it, or introduce a systematic error compensation for the whole sensor batch or production. After removing the systematic error, the remaining random errors can be calibrated per individual sensor, so that ultimately the error distribution fits within the accuracy limits. Fig. 3-2A may represent the statistical distribution of the offset for a large batch of sensors, showing an average (μ) of 45 mV and a standard deviation (σ) of 5 mV. For the example, we suppose that the desired accuracy allows a maximum offset of ±2 mV.

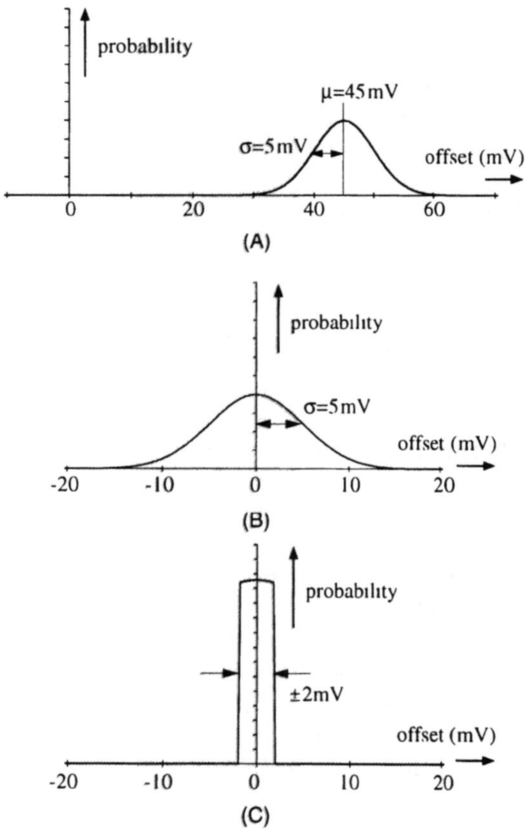

Fig. 3-2 *Statistical offset distribution, before (A), and after systematic correction (B), and calibration (C).*

Practically all sensors (99.76%) have an offset within the $\mu \pm 3\sigma$ range, and so, initially, none of them has an offset smaller than 2 mV. To calibrate the offset error, a correction of maximum 60 mV with a 4 mV accuracy would be required. By applying a systematic offset correction of –45 mV first, the shifted distribution of Fig. 3-2B is obtained, and approximately 30% of the sensors have an offset according to the specification. Next, an offset calibration with a correction range of ±15 mV and a resolution of 4 mV (~3-bit) is sufficient to put almost all sensors in the ±2 mV offset range. The offset error distribution of the sensors, after such a calibration is shown in Fig. 3-2C. The calibration accuracy can be increased to further reduce the offset error and the width of the distribution.

3.1.3 Linear and nonlinear errors

We can distinguish between linear errors and nonlinear errors. Offset and gain errors and their first-order temperature coefficients are linear errors.

In order to correct linear errors only linear operations are required, which are easy to implement in circuit technology. In a straightforward manner a programmable offset and programmable gain can be realized. Some examples will be given at the beginning of *chapter 4*. The calibration or correction of the nonlinearity, referred to as linearization, is more complicated. We need to counteract the sensor's nonlinearity and somehow realize an inversely nonlinear electronic transfer. Many different methods can be found in literature. The basic linearization principles are explained in section 3.2. In section 3.3 we propose and explain the polynomial linearization method which will be used in the realizations presented in subsequent chapters.

For linearity errors we should distinguish also between systematic and random errors. A systematic nonlinearity, which can be explained from the sensor's physical model or reproduced consistently from measurements, should also be corrected systematically. An example of a systematic linearity compensation will be given in section 3.2.1. Preferably, calibration is used only to compensate random errors resulting from device-to-device variations. A certain number of measurements need to be taken in order to determine and correct a sensor's nonlinearity. The number of measurements necessary to obtain a certain reduction of the linearity error, strongly depends on the linearization method used. To reduce the costs of calibration it is important to minimize the number of measurements, because measurements take (production) time. This is an important criterium in the selection of an appropriate linearization method for sensor calibration.

3.1.4 Normalized sensor transfer functions

One-dimensional sensor transfer function

To simplify the mathematical descriptions of the linearization methods we will work with normalized sensor functions. The electrical output signal e_{out} of a sensor can normally be considered as a function S of one physical input signal φ_{in}: $e_{out} = S(\varphi_{in})$.

Through normalization, the range of interest $[\varphi_{min}, \varphi_{max}]$ for the physical input signal is mapped on the range [-1, 1] for a normalized input signal. Similarly, the desired output range $[e_{min}, e_{max}]$ of the electrical sensor output must be mapped on the [-1, 1] range for a normalized output signal.

We can derive the *normalized input signal* and *normalized output signal* of the sensor respectively as:

$$x = \frac{2\varphi - \varphi_{max} - \varphi_{min}}{\varphi_{max} - \varphi_{min}} \qquad y = \frac{2e - e_{max} - e_{min}}{e_{max} - e_{min}} \tag{3-1}$$

The normalized sensor output y can be described by the *normalized transfer function*, which is expressed as a one-dimensional function of the normalized input signal x: $y = f(x)$. Furthermore, we assume that the desired normalized transfer is a linear function of x with unity gain: $y = g(x) = x$.

A normalized transfer function example, $f(x)$, and the ideal transfer function $g(x)$ are shown in Fig.3-3. We will use the example curve for the explanation of the linearization methods in the next section. The shown function $f(x)=-1.2 \bullet ln(1.7-x)$ is taken as an arbitrary example, and not because it would characterize a typical sensor. Assuming that for a real case sensor the systematic errors have been compensated first, then only arbitrary errors would remain. Thus we do not know beforehand how to describe the transfer function $f(x)$. Instead, we have to take several, say N, calibration measurements $f(x_n)$ [n=1 to N] which can be used to compose an interpolation function $f'(x)$ which resembles the real sensor transfer. On the basis of the interpolation function, we can linearize the sensor transfer so that it approaches the ideal sensor curve $g(x)$. The linearized transfer

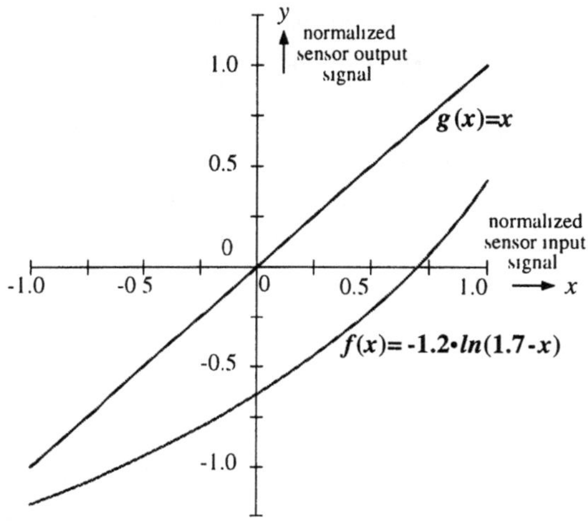

Fig. 3-3 *Normalized one-dimensional sensor transfer function, and ideal transfer function.*

function $y=h(x)$ will thus be composed of the sensor output signal $f(x)$, the calibration measurements $f(x_n)$, and the corresponding ideal output values $g(x_n)$:

$$h(x) = H(f(x), f(x_n), g(x_n)) \tag{3-2}$$

The obtained linearization can best be examined by looking at the error function, defined as:

$$\varepsilon(x) = h(x) - g(x) \tag{3-3}$$

Because of the normalization, the error curve can also be directly interpreted as the relative error (with respect to one half of the symmetrical scale).

Two-dimensional sensor transfer function

If we want to examine sensors with cross-sensitivity to another physical input signal, we must consider the sensor transfer as a two-dimensional function. We may express the electrical sensor output e_{out} as a function S of two physical input signals φ_{in} and ψ_{in}: $e_{out} = S(\varphi_{in}, \psi_{in})$. Again, we may transform the electrical output signal to a normalized output y, and the physical input variables to normalized input signals x and z. We can then examine the normalized two-dimensional transfer function $y = f(x, z)$. Ideally, the normalized output y should be linearly dependent on the input variable x with unity gain, but independent of the other input z (zero-gain). Thus the ideal normalized function is given by: $y = g(x, z) = x$. The two-dimensional transfer functions can be presented graphically as a surface in a three-dimensional (3D) picture. Examples are shown in Fig. 3-4. In this case, the goal of the calibration is to process the sensor

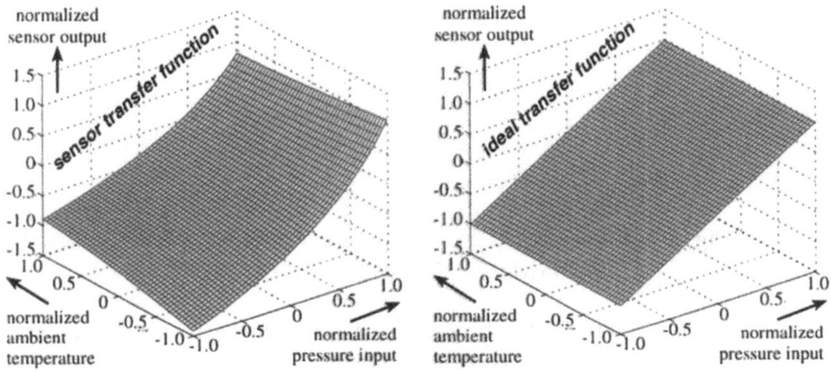

Fig. 3-4 *Normalized two-dimensional sensor transfer function, and ideal sensor transfer function.*

transfer function $f(x,z)$ in such a way that a function $h(x,z)$ results, which resembles the linear function $g(x,z)$. This will require calibration measurements $f(x_n,z_m)$ for various input conditions defined by x_n [n=1 to N] and z_m [m=1 to M]. Referring to the number of measurements this is called an $N \times M$ calibration. Similarly to equation (3-3), we can define the two-dimensional error surface $\varepsilon(x, z) = h(x, z) - g(x, z)$.

3.2 Linearization

Some types of sensors show a characteristic nonlinear transfer, which can be explained from a physical model, or which can be reproduced consistently. In that case, the required calibration can be simplified by first applying a systematic linearization, as proposed in section 3.2.1. The remaining random variations of the nonlinearity can then be calibrated by using a general linearization method. Such methods are applied directly for sensors with a more or less linear transfer, showing only linearity errors that vary randomly from device-to-device. The following sections will explain several such linearization methods. All those methods are based on the use of calibration measurements of the sensor output signal. Besides for linearizing the signal transfer, the calibration measurements are also used for correcting the offset and gain error.

3.2.1 Sensor characteristic linearization

Certain kinds of sensors always show a typical nonlinear transfer curve. The shape of the nonlinearity is reproducible and can often be traced in the model of the sensor. Such a systematic error should preferably be corrected by a systematic linearization, rather than by means of interpolation of multiple calibration measurements. A sensor may, for example, typically show a logarithmic transfer which could be written in the form of equation (3-4) with *a*, *b*, *c*, and *d* as sensor-dependent constants:

$$e_s = f(\varphi) = a + b \cdot \ln(c + d \cdot \varphi) \tag{3-4}$$

Such a logarithmic transfer curve can easily be linearized by using an electronic circuit with an exponential transfer function. We could use the exponential relation between base-emitter voltage and collector current of a bipolar transistor [3], see equation (2-1). The idea is illustrated in Fig. 3-5. The sensor output signal e_s is applied at the base-emitter voltage. The collector current I_c is converted into a output voltage V_{out} by an IV-converter. For the overall transfer of sensor and circuit, we achieve:

$$\begin{aligned} V_{out} &= h(\varphi) = R \cdot I_C = R \cdot e^{K_T V_{be}} \\ &= R \cdot e^{K_T \{a + b \; \ln(c + d \cdot \varphi)\}} \\ &= R \cdot e^{aK_T} \cdot e^{bK_T} \cdot (c + d \cdot \varphi) \end{aligned} \tag{3-5}$$

which can be written in a simplified form as:

$$h(x) = A + B \cdot x \tag{3-6}$$

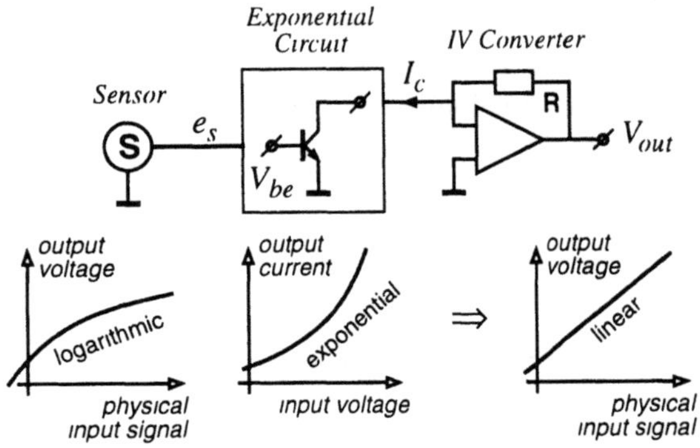

Fig. 3-5 *Linearization of a typical logarithmic transfer function.*

Due to the inverse transfer curve of the transistor with respect to the transfer of the sensor, the output is now linearly proportional to the physical input signal φ, and only the gain and offset errors remain to be calibrated. Attention should be paid to the temperature sensitivity introduced by $K_T = q/kT$. This temperature sensitivity could be cancelled easily by multiplying the sensor signal e_s with a PTAT (proportional to *T*) signal, before applying it to the transistor [3].

It may not always be possible to find a way to integrate the inverse function of the sensor curve in an electronic circuit. Furthermore, the compensation applied may not sufficiently reduce the nonlinearity. In those cases, additional linearization can be improved, using one of the approaches presented in the next sections.

3.2.2 Linearization based on a look-up table

A simple approach to correct any type of transfer is to store the complete inverse function of the sensor transfer in a *table* and to *look up* the corrected output value corresponding to the sensor output signal measured [4,5,6,7]. An Analog-to-Digital converter (ADC) is used to quantize the sensor output signal into a digital number, which can be used as an address of the memory containing the look-up table. At the memory address, the corrected digital value for the desired transfer curve can be read out. The digital output can be converted into an analog signal by means of a Digital-to-Analog converter (DAC). A graphical explanation is shown in Fig. 3-6.

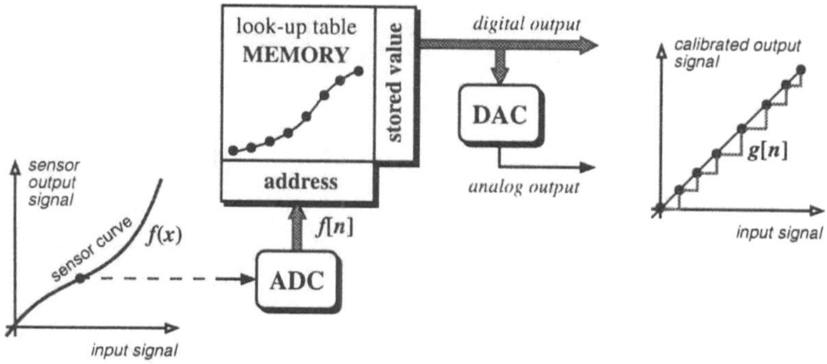

Fig. 3-6 *Look-up table calibration method.*

During the calibration phase, the correct output signals must be stored in the memory at the correct addresses. This can be done by (slowly) scanning through the range of the physical input signal of the sensor, and storing the desired digital output values in the memory, at each single address change. The correct digital output values could be obtained from a reference sensor with an ADC [6]. Filling the complete look-up table requires a large series or 'scan' of calibration measurements. Another approach is to limit the calibration to a few measurements, and use an advanced interpolation algorithm on a computer to obtain the required table values.

The quantization of the uncorrected nonlinear sensor output signal implies a very uneconomical use of the ADC's dynamic range. First of all, when large offset and gain errors (distributions) are taken into account, it is possible that some sensors will use only a small fraction of the ADC's input range. Furthermore, a nonlinear transfer may locally show a low sensitivity, which means less ADC quantization levels will be available for a certain range of input signals. This effect will become clear in the example, shown in Fig.3-7.

A 6-bit quantization of the normalized input signal over the [-2,2] range was used, and applied to the example sensor function $f(x)$. The quantized signal $f_{LUT}(x)$ is shown in Fig.3-7. Each quantization level on the y-axis can be used to obtain a corrected output value, which results in the calibrated signal transfer function $h_{LUT}(x)$. It is clear that for input signals x in the range [-1,0], there are less quantization levels available than in the [0,1] range due to the nonlinearity of $f(x)$. The corresponding error curve,

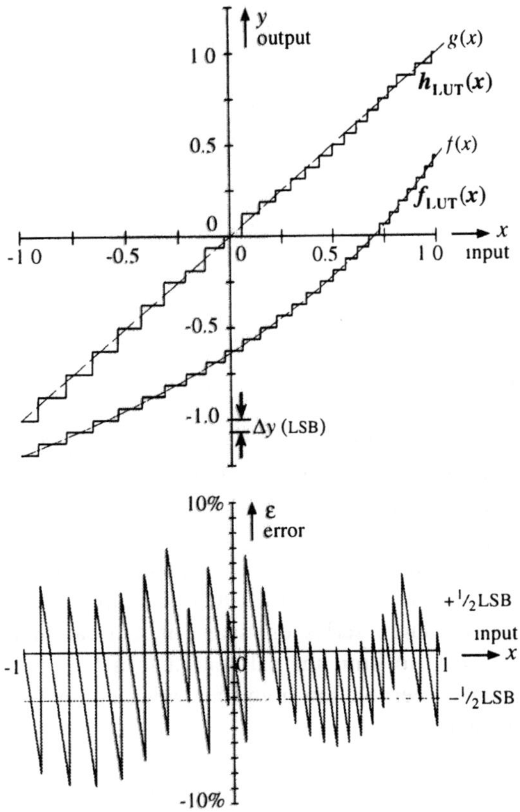

Fig. 3-7 *Look-up table linearization of the transfer curve, and corresponding error curve.*

also shown in Fig. 3-7, demonstrates that the quantization error will not remain within the $\pm{}^1/_2$ LSB limits.

The disadvantages of using a look-up table can be listed as:

- *requirement of a large memory: for an 8-bit quantization 256 bytes are needed, for 16-bit 135* **kilo***bytes are needed*
- *many calibration measurements needed or a good interpolation to obtain all values for the look-up table*
- *quantization of the (nonlinear) sensor output signal is needed*

Advantages of the method can be summarized as:

+ *once the look-up table is filled, a very fast correction of the sensor signal is possible, as no signal processing is needed*
+ *the look-up table calibration can be realized in a simple straight-forward hardware or software implementation*
+ *the calibration includes the errors of the sensor interface circuit and the AD-converter*

The system configuration shown in Fig. 3-6 can easily be expanded to achieve calibration of a two-dimensional sensor transfer curve. In case of cross-sensitivity to temperature, we can add a temperature sensor with an ADC. The digital temperature signal is then used in combination with the original sensor signal to compose an address for the look-up table. Naturally, memory requirements and the number of calibration measurements will increase accordingly.

3.2.3 Linearization based on piecewise-linear interpolation

Instead of storing the complete sensor transfer curve in a large memory, we could use a small memory and store only the coefficients of a model which describes the sensor transfer. We can measure the sensor output signals for a small set of known input signals. These calibration measurements represent 'knots' on the sensor transfer curve. Between all adjacent knots we can draw straight lines, described by $y = a_n + b_n x$. These lines compose the piecewise-linear (PWL) interpolation $f_{pwl}(x)$ of the sensor transfer curve. Based on these linear 'portions', the sensor output signal can be processed in a simple way to obtain a linearized transfer curve. In each sub-range an 'offset' and a 'gain' correction will be applied. Only the values indicating the sub-ranges and the required offset and gain corrections for each sub-range need to be stored in a memory.

The method is explained graphically in Fig. 3-8. After the piecewise-linear model of the sensor transfer has been established in the calibration phase, a range detection is sufficient to determine what offset and gain correction must be applied to the sensor signal. Although a digital implementation is the most obvious, the signal processing can also be done in the analog signal domain. Hence the dotted print used for the ADC and the DAC in the picture of Fig. 3-8.

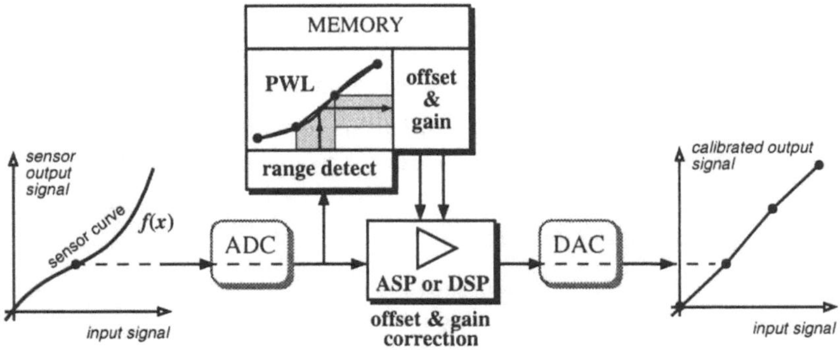

Fig. 3-8 *Piece-wise-linear method for linearity calibration.*

A piecewise-linear interpolation of the example transfer curve is shown in Fig. 3-9. It is based on five calibration measurements represented by the knots (x_1,y_1) to (x_5,y_5). The effect of quantization is not taken into account as it is not essential for the piecewise linearization to quantize the sensor output signal.

The linear interpolation between the first two knots can be found as:

$$f_1(x) = f(x_1) + \frac{f(x_2) - f(x_1)}{x_2 - x_1} \cdot (x - x_1) \qquad (3\text{-}7)$$

Which can be proved simply by verifications with $x=x_1$ and $x=x_2$.

With N knots the complete PWL curve can be described mathematically by the sum of the $N-1$ linear components $f_n(x)$:

$$\begin{aligned} x \in [x_n, x_{n+1}) &\rightarrow f_n(x) = f(x_n) + \frac{f(x_{n+1}) - f(x_n)}{x_{n+1} - x_n}(x - x_n) \\ x \notin [x_n, x_{n+1}) &\rightarrow f_n(x) = 0 \end{aligned} \qquad (3\text{-}8)$$

$$f_{pwl}(x) = \sum_{n=1}^{N-1} f_n(x) \qquad (3\text{-}9)$$

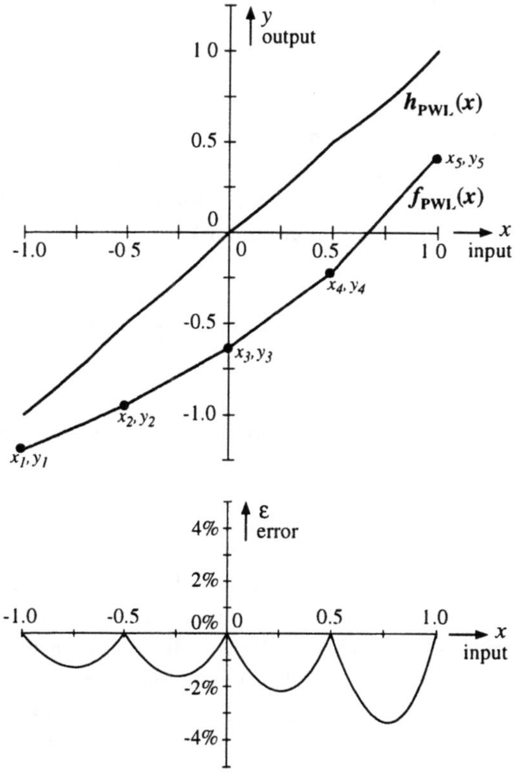

Fig. 3-9 *Piecewise-linear interpolation, and corresponding linearization and error curve.*

We are not really interested in the *interpolation* curve of the sensor transfer, but rather obtain an expression for the *linearization* function of the sensor transfer. Similar to equation (3-8) we can describe the components $h_n(x)$ which transform the output signal $f(x)$ as follows:

$$f(x) \in [y_n, y_{n+1}) \rightarrow$$

$$h_n(x) = g(x_n) + \frac{g(x_{n+1}) - g(x_n)}{y_{n+1} - y_n} \cdot \{f(x) - y_n\} \quad (3\text{-}10)$$

$$f(x) \notin [y_n, y_{n+1}) \rightarrow \qquad h_n(x) = 0$$

based on the desired output values $g(x_n)$ and the measured output signals:

$$y_n = f(x_n) \tag{3-11}$$

By substitution it can be proven that the values of components $h_n(x)$ match the desired output values in two calibration points, $h_n(x_n)=g(x_n)$ and $h_n(x_{n+1})=g(x_{n+1})$. The linearization function $h_{PWL}(x)$ is then defined as the sum of the sub-range components:

$$h_{pwl}(x) = \sum_{n=1}^{N-1} h_n(x) , \tag{3-12}$$

and thus intersects the desired transfer function $g(x)$ at all input values x_n used for the calibration.

The resulting calibrated transfer curve and the error curve for the example case with five calibration measurements are shown in Fig.3-9. The linearity peak error is reduced to 5%, while the uncalibrated nonlinearity was about 30%. The error reduction could be improved somewhat by optimizing the division of the sub-ranges, instead of taking equidistant calibration points x_n. This would only be possible for specific transfer curves of which the shape is known beforehand. The linearization can be further improved by increasing the number of interpolation knots, at the expense of more calibration measurements and a larger memory requirement. Alternatively, a better interpolation between the knots of the transfer curve can be used to improve the linearization, as will be explained in the next section.

The advantages and the disadvantages of the PWL calibration and linearization method can be summarized as:

+ *low memory requirements*
+ *a reasonably small number of calibration measurements is required*
+ *fast and simple signal correction; no advanced signal processing is required, only gain and offset correction*
- *reduction of the linearity error is limited; very nonlinear transfer curves still require a large number of calibration steps*

- *separate calculation of offset and gain correction is required for each subsection of the signal range*
- *large discontinuities in the derivative of the linearized signal*

3.2.4 Linearization based on piecewise-polynomial or spline interpolation

Rather than using linear interpolating between two knots on the sensor transfer curve, we can use a polynomial interpolation between three or more consecutive knots. These interpolations can also be expressed mathematically as splines [8,9]. A series of such interpolations or splines can be used to compose a 'piecewise-polynomial' or spline function which closely resembles the original sensor transfer function over the complete input range [10,11,12].

The second-order polynomial interpolation through three knots can be described by:

$$f_n(x) = f(x_n) + a_n \cdot (x - x_n) + b_n \cdot (x - x_n)(x - x_{n+1}) \quad (3\text{-}13)$$

with the coefficients a_n and b_n based on the three measurements $y_n = f(x_n)$, $y_{n+1} = f(x_{n+1})$, and $y_{n+2} = f(x_{n+2})$ of the transfer function:

$$a_n = \frac{y_{n+1} - y_n}{x_{n+1} - x_n} \quad (3\text{-}14)$$

$$b_n = \frac{y_{n+2} - y_n}{(x_{n+2} - x_n)(x_{n+2} - x_{n+1})} - \frac{y_{n+1} - y_n}{(x_{n+1} - x_n)(x_{n+2} - x_{n+1})}$$

We can divide the input signal range into sub-ranges, based on the measurement points x_n. In each sub-range there are two polynomial interpolations available, except for the first and the last sub-ranges where just one interpolation is valid. Therefore, we define the linearization function $f_{SPL}(x)$ for N interpolation knots as the following combination of the interpolation components or splines $f_n(x)$:

$$\begin{aligned} &x \in [x_1, x_2) \rightarrow f_{spl}(x) = f_1(x) \\ &x \in [x_n, x_{n+1})\Big|_{n=2}^{N-2} \rightarrow f_{spl}(x) = \frac{f_n(x) + f_{n+1}(x)}{2} \\ &x \in [x_{N-1}, x_N] \rightarrow f_{spl}(x) = f_{N-1}(x) \end{aligned} \tag{3-15}$$

Thus, in the sub-ranges where there are more interpolation functions available they are averaged.

Equations (3-13) to (3-15) describe the *interpolation* function of the sensor transfer curve $f(x)$ and *not* the *linearization* function. We can apply the same approach as was used for the piecewise-linear case to find the demanded linearization function $h_{SPL}(x)$.

First, the sub-range components are described by:

$$h_n(x) = g(x_n) + A_n\{f(x) - y_n\} + B_n\{f(x) - y_n\}\{f(x) - y_{n+1}\} \tag{3-16}$$

in which the coefficients A_n and B_n can be based on the calibration measurements of the transfer function, indicated by $y_n = f(x_n)$, as follows:

$$\begin{aligned} A_n &= \frac{g(x_{n+1}) - g(x_n)}{y_{n+1} - y_n} \\ B_n &= \frac{g(x_{n+2}) - g(x_n)}{(y_{n+2} - y_n)(y_{n+2} - y_{n+1})} - \frac{g(x_{n+1}) - g(x_n)}{(y_{n+1} - y_n)(y_{n+2} - y_{n+1})} \end{aligned} \tag{3-17}$$

The linearization function $h_{SPL}(x)$ is then composed of the components $h_n(x)$ in the same way as the interpolation function in equation (3-13).

Fig.3-10 shows the interpolation and linearization function for the example case based on five calibration points (x_1,y_1) to (x_5,y_5), and also the resulting error curve.

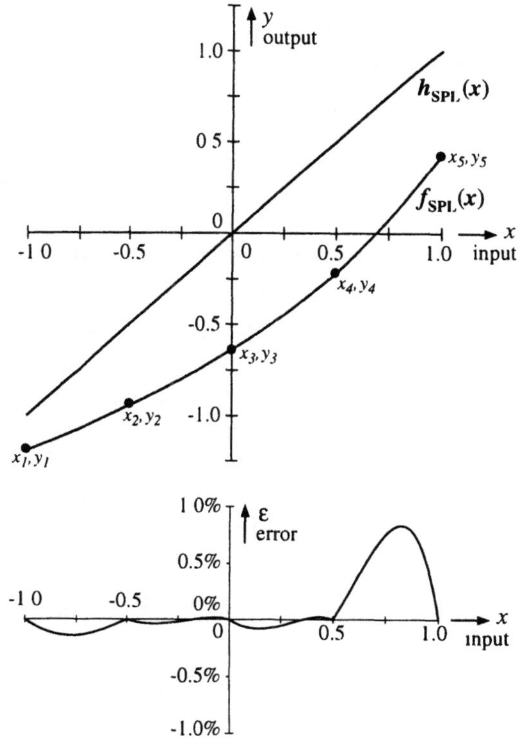

Fig. 3-10 *Spline interpolation and linearization functions, and corresponding error curve.*

Clearly, the linearity error is reduced significantly from about 30% for the uncalibrated transfer function to below 1% for the linearized function. Here again, we see that the peak error is in the x=0.5 to x=1 range and could have been reduced by a better choice of the calibration points x_n. For finding the optimal choice based on statistical information about the sensor function, Crary even developed an advanced computer program [13, 14]. Again, it only helps for transfer curves with a clearly characteristic nonlinearity, and a limited deviation from this curve for a batch of sensors.

Advantages and disadvantages of the spline interpolation technique are:

+ *good linearization can be obtained with a low number of calibration measurements*

- + *systematic approach is used, which can be set up with recursive equations [9] very suitable for software implementations*
- – *all calibration measurements need to be executed and stored before correction curves are computed*
- – *for each sub-range a different correction function is computed, using different coefficients*
- – *advanced computations are needed, preferably implemented in computer or microcontroller software*

Further improvement in the linearization can be expected when even more knots are used for each interpolation, or, in other words, when higher degree splines are used [9]. When all measurements are used in one single interpolation function we obtain an expression of a curve that 'fits' the entire sensor function on the full input range. Similarly, we can obtain an inverse interpolation curve which linearizes the sensor transfer function. This case will be described in the next sections.

3.2.5 Linearization based on curve fitting

As explained in section 3.2.1 it is possible to counteract the nonlinear transfer of the sensor with an inversely nonlinear transfer when the expression for the sensor transfer curve is known. It is also possible when the sensor transfer function is not precisely known but only characterized by some calibration measurements [1,2,4,5,15,16,17]. Curve-fitting techniques make it possible to compute a mathematical function which intersects the actual sensor function in the measurement points. This is usually done by defining a weighted sum of expressions based on the sensor output $y=f(x)$. The weight factors are calculated on the basis of the measurements $f(x_n)$ of the sensor transfer function. One may then try to compute the exact *inverse function* of the found curve-fit function. A better approach is to compose a curve-fit function $H\{f(x)\}$ for the inverse sensor transfer right away. The curve fitting then refers to matching the resulting transfer function $h(x)=H\{f(x)\}$ to the desired linear transfer function $g(x)$. The linearization function can usually be expressed in the following form:

$$h(x) = \sum_{n=0}^{N} c_n \cdot E_n\{f(x)\} \qquad (3\text{-}18)$$

The expression E_n depends on the specific curve-fit technique used, and could be, for example, an exponential, sinusoidal, or polynomial function of increasing order (n). The weight factors c_n, also referred to as calibration coefficients, are calculated, using the calibration measurements $f(x_n)$ and the corresponding values for the desired transfer $g(x_n)$.

The computations required can be split in two groups. First, the calculation of the correct values for calibration coefficients during or after the calibration phase. This may be done externally of the sensor (on a computer). Second, calculation of the linearized output signal based on the sensor output and the calibration coefficients. This correction must be integrated with the smart sensor, either in analog or digital hardware, or in software in a microcontroller interfacing the sensor. Memory must be present to store the calibration coefficients. The procedure is explained graphically in Fig. 3-11.

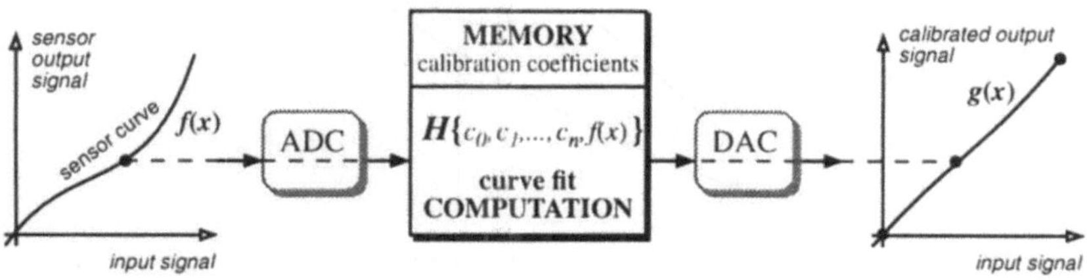

Fig. 3-11 *Inverse transfer function method for linearity calibration.*

A well-known curve-fitting technique makes use of polynomial functions and Lagrange interpolation. For N calibration measurements, the linearization function can be described by:

$$h(x) = c_0 + c_1 (f(x))^1 + c_2 (f(x))^2 \ldots + c_{N-1} (f(x))^{N-1} \tag{3-19}$$

And the N coefficients can be calculated by taking the *complete* set of calibration measurements:

$$c_n = k(x_1, y_1, \ldots, x_N, y_N), \tag{3-20}$$

which will not be worked out here. A modified polynomial calibration technique will be elaborated in section 3.3.

As an example of the curve-fitting technique, Fig. 3-12 shows the linearized transfer curve $h(x)$ based on a fourth-order polynomial of the sensor function $f(x)$. The calculation of the coefficients then requires five calibration measurements. The corresponding error curve shows how well the method reduces the nonlinearity.

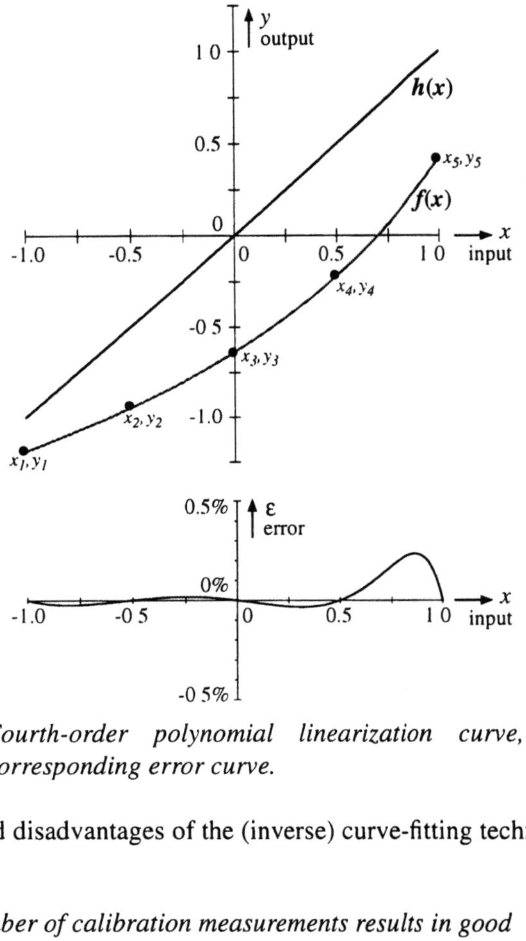

Fig. 3-12 *Fourth-order polynomial linearization curve, and corresponding error curve.*

Advantages and disadvantages of the (inverse) curve-fitting technique can be listed as:

+ *low number of calibration measurements results in good linearization*
+ *low memory requirements because of a low number of coefficients*
+ *one single correction formula for the complete signal range*

- *higher order polynomials might require high-accuracy computations (floating-point operations)*
- *rather complex calculation of calibration coefficients may be needed*

This curve-fitting technique based on polynomial interpolation can be set up in a such a way that is compatible with the spline function method. The main difference is that the spline method uses multiple polynomial interpolations over sub-ranges with a subset of the measurement points, while the curve-fit polynomial uses one interpolation over the full range, using all measurement points.

3.2.6 Linearization based on error minimization

Another interpolation technique, much different from the previous ones, is based on constructing and *tuning* a curved function which does not exactly go through all the measured data points, but which still approaches the measured curve closely. Using a mathematical algorithm, the coefficients of the interpolation curve are tuned in such a way that the errors at the measured data points (difference between measured y-value and y-value on the interpolation curve) are minimized. Usually the *mean square error* is taken as the error function that should be minimized. Such methods can be applied provided that there is some sort of redundancy in the calibration data, - we have a-priori knowledge of the measured curve (a model) and we have more than the minimum number of calibration measurements required to compose a corresponding interpolation curve.

Suppose, for example, that we know that the measured transfer should obey a quadratic function and that we have four measurement points. Then, it is better to tune the three coefficients of a second order polynomial (quadratic curve) and *minimize* the errors in the four points, rather than to calculate a third order polynomial and *zero* the errors. The reason for this is that the last option might introduce a much larger error in *between* the measured data points, as we knew before-hand that the introduced third order polynomial term should be zero. However, based on the a-priori knowledge we could also conclude that have taken one unnecessary calibration measurement, as three measurement points would have been sufficient to compose a second order interpolation curve intersecting the measured curve.

The technique for interpolation explained above can also be applied for linearization. While the linearization techniques presented in the previous sections required straight-forward calculations only, the optimization of several (or many) coefficients for the minimization of the error function requires a different class of mathematics [13,14]. How to determine the optimum order for the linearization curve? Where to start tuning the calibration coefficients? How to determine a good error function and how to minimize it and assure a global (rather than a local) minimum? These are complex issues, especially for higher order linearization and two-dimensional calibrations. Solutions of such optimization problems are usually found by using iterative algorithms (e.g. simulated annealing) implemented in software [13,14]. Considering the amount of complex calculations required, the approach is not very suitable for realization in hardware integrated on a smart sensor[1], nor for an elaborate treatise fitting within the scope of this book.

We confine ourselves to summarizing the pro's and con's of linearization techniques based on error minimization, but further ignore the option:

- \+ *ideally suited for using large numbers of calibration measurements, or*
- \+ *effectively using redundant information or a-priori knowledge about the sensor transfer function*
- \+ *good linearization with a very low error can be obtained, provided the correct choice of a linearization function based on a known or assumed sensor model*
- \+ *good alternative approach for when increasing the order of a (polynomial) curve fitting does not improve the error.*
- \- *storage of the (many) calibration measurements is needed, because the are used in the error function optimization*
- \- *complex iterative calculations required for finding the optimum choice of calibration coefficients*
- \- *requires redundant information, and thus more than the minimum number of calibration coefficients*
- \- *unsuitable for hardware implementation*

[1] However, it is not unthinkable that a simple linearization/calibration function is integrated in the sensor hardware, where as the calculation of the optimal calibration coefficients is done externally in computer software.

3.2.7 Conclusion

Different linearization techniques were presented, which can be used to calibrate the sensor transfer function. Sensor calibration techniques found in literature are practically all based on one of these techniques or on a combination of them [18,19]. Hardware and software implementations of the look-up table and the piecewise-linear techniques can be found, whereas the more advanced techniques are always implemented in software. Application of sensor characteristic linearity compensation in analog electronic circuitry can be found for thermal sensors especially [20,21].

With respect to the complexity of the hardware that needs to be integrated with the sensor, the look-up table and piecewise-linear methods are favored; the software implementations can be (hybridly) integrated in the form of a microcontroller. However, using a look-up table, it is required to vary the physical input signal step-wise through the whole signal range. This will be time-consuming for physical signals, such as temperature and pressure, having a slow response or large settling time (which depends on the actuator used). Dynamic scanning in combination with a reference sensor may sometimes be used to speed up the look-up table calibration [6].

Furthermore, it is made clear that, with respect to the number of calibration measurements, the best linearization can be achieved with the spline interpolation and the curve-fitting techniques.

In all cases, a set of calibration measurements has to be completed and stored first. These will then be used to compute the coefficients needed in the correction of the sensor signal. A disadvantage of that approach is that we have to choose the number of calibration measurements beforehand. If the calibration appears to reduce linearity errors insufficiently, we have to execute additional calibration measurements or a new and larger set of measurements, and completely *redo* the calibration calculations. Using the polynomial (curve-fit) calibration technique that will be explained in the next section, such recalculations can be avoided.

3.3 Progressive polynomial calibration method

The polynomial calibration and linearization method we propose operates on the principle that each calibration measurement can be used directly to calculate one programmable coefficient in the correction function. The correction is then immediately applied to modify the sensor output. The next calibration step makes use of this corrected sensor signal. Each succeeding correction step is applied in such a way that the previous calibrations remain undisturbed.

The first measurement is done to calibrate the offset which means the transfer function is shifted or *translated*, as shown in the first picture of Fig.3-13.

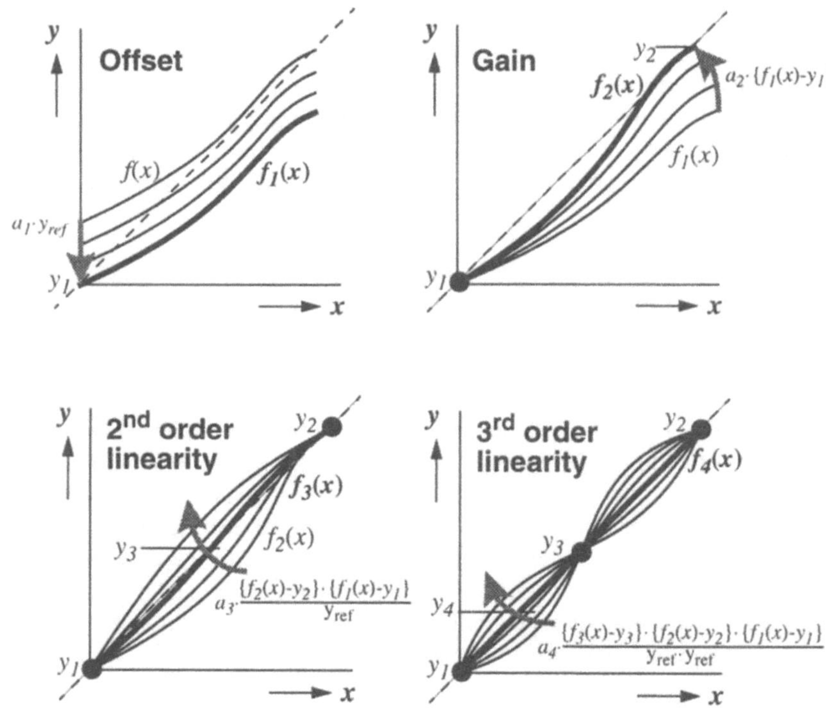

Fig. 3-13 *Four steps of the progressive polynomial calibration/ linearization method.*

A following calibration measurement is used to correct the gain error, without affecting the offset calibration. This is achieved by *rotating* the function around the previous calibration point, as shown in the second picture of Fig.3-13. In this way the previous offset correction at $y=y_1$ will not be affected by the gain correction. The cross-dependence between offset and gain setting, found in most other calibration methods, is eliminated in the calibration method proposed here.

The next calibration measurements are used to correct linearity, which is achieved by '*bending*' the function in such a way that the previous calibration points stay fixed. The third picture in Fig.3-13 shows the second order linearity correction, and the last picture of Fig.3-13 shows the third order linearization. More calibrations can be done in the same manner to further linearize the sensor transfer function. This explains an incremental way of constructing a polynomial correction curve to fit the desired linear transfer function. In fact, at each calibration step, the calibrated transfer function progresses towards the desired transfer, hence the label *progressive polynomial calibration*.

We will now describe the mathematical aspects of this calibration method.

3.3.1 One-dimensional functions

In accordance with section 3.2, the (physical) input variable is indicated as x, the (electrical) sensor output is indicated by y; the uncalibrated sensor response is denoted by the sensor transfer function $y=f(x)$; the desired transfer function is denoted as $y=g(x)$ and is assumed to be a linear function of the input signal $g(x)=K\cdot x$. The purpose of the calibration is to obtain a calibrated sensor response function which closely resembles the desired transfer function. This is done by taking calibration measurements for a set of well-known input signals x_n, and comparing the measured sensor output value $f(x_n)$ to the desired output value $y_n=g(x_n)$. The proposed approach is to calculate a corrected sensor transfer curve $h_n(x)$ after *each* calibration measurement. We make use of the previously corrected transfer curves $h_1(x)$ to $h_{n-1}(x)$, the previous calibrated output values y_1 to y_{n-1}, and the n-th calibration measurement $f(x_n)$. The calibration is thus expressed in a *series of nested formulas*:

$$h_n(a_n, h_1(x)...h_{n-1}(x), y_1...y_{n-1}) \tag{3-21}$$

in which a_n is referred to as the n-th *calibration coefficient*. At each calibration step the calibration coefficient a_n has to be calculated in order to obtain that the corrected sensor output becomes equal to the desired output: $h_n(x_n)=g(x_n)=y_n$.

We will clarify the procedure by showing the first four steps.

- The first calibration function $h_1(x)$ is given by:

$$h_1(x) = f(x) + a_1 \cdot y_{ref} \tag{3-22}$$

A (constant) reference signal y_{ref} is needed only to give the correction factor the same unit as the functions, so that a_1 can be a dimensionless number[1]. Based on the first calibration measurement $f(x_1)$ and the desired output value $y_1=g(x_1)$, the calibration coefficient a_1 can be calculated:

$$a_1 = \frac{y_1 - f(x_1)}{y_{ref}} \tag{3-23}$$

Substitution of equation (3-23) into equation (3-22) proves that the desired equalization $h_1(x_1)=y_1$ is obtained.

- The second calibration function $h_2(x)$ uses the previously corrected function $h_1(x)$ as follows:

$$h_2(x) = h_1(x) + a_2 \cdot \{h_1(x) - y_1\} \tag{3-24}$$

The additive correction term is zero in the first calibration point (x_1,y_1), which keeps the previous equalization still valid for the new function $h_2(x_1)=h_1(x_1)=y_1$, independent of a_2. The second equalization $h_2(x_2)=y_2$ has to be obtained by finding the correct value for calibration coefficient a_2. This requires a second calibration measurement $f(x_2)$, from which

[1] The reference signal is introduced for compatibility with the implementation in *chapter 4*. Normalization of the input and output signals, see page 56, can still be applied, but is not necessary.

$h_1(x_2)$ can be calculated, and can then be compared with the desired output value $y_2=g(x_2)$. The equalization is obtained when:

$$a_2 = \frac{y_2 - h_1(x_2)}{\{h_1(x_2) - y_1\}} \qquad (3\text{-}25)$$

• The third calibration function should be constructed in such a way that the previous two equalizations remain valid, while a second-order polynomial correction is obtained. A correction term that is zero in the previous two calibration points (x_1,y_1) and (x_2,y_2) is added in the third calibration function:

$$h_3(x) = h_2(x) + a_3 \cdot \frac{\{h_1(x) - y_1\} \cdot \{h_2(x) - y_2\}}{y_{ref}}, \qquad (3\text{-}26)$$

where a reference y_{ref} is only inserted to preserve the unit of the correction term. A third calibration measurement $f(x_3)$ is done and immediately corrected by calculating $h_1(x_3)$ and $h_2(x_3)$, which are used to find the desired value for the third calibration coefficient a_3:

$$a_3 = \frac{\{y_3 - h_2(x_3)\} \cdot y_{ref}}{\{h_1(x_3) - y_1\} \cdot \{h_2(x_3) - y_2\}} \qquad (3\text{-}27)$$

Substitution into equation (3-26) proves that the a third equalization of the calibrated sensor transfer $h_3(x)$ to the desired transfer $g(x)$ has been achieved, namely $h_3(x_3)=y_3$.

• Similarly a third order polynomial correction can be realized in the fourth calibration function, which is formulated as:

$$h_4(x) = h_3(x) + a_4 \cdot \frac{\{h_1(x) - y_1\} \cdot \{h_2(x) - y_2\} \cdot \{h_3(x) - y_3\}}{y_{ref} \cdot y_{ref}} \qquad (3\text{-}28)$$

Here too, the reference y_{ref} is used to preserve the unit of the correction term. Again a new calibration measurement $f(x_4)$ is needed to obtain

$h_1(x_4)$, $h_2(x_4)$, and $h_3(x_4)$, which enables the calculation of the fourth calibration coefficient a_4:

$$a_4 = \frac{\{y_4 - h_3(x_4)\} \cdot y_{ref}^2}{\{h_1(x_4) - y_1\} \cdot \{h_2(x_4) - y_2\} \cdot \{h_3(x_4) - y_3\}} \tag{3-29}$$

After four calibration steps, a calibration function $y=h_4(x)$ is obtained which intersects the desired transfer function $y=g(x)$ in the four chosen calibration points (x_1,y_1), (x_2,y_2), (x_3,y_3), and (x_4,y_4). In fact, we have been building up a *Lagrange* interpolation polynomial for the sensor transfer function. Equations 3-22, 3-24, 3-26, and 3-28 can be worked out to obtain the standard polynomial form:

$$h_4(x) = c_0 + c_1 \cdot f(x) + c_2 \cdot \{f(x)\}^2 + c_3 \cdot \{f(x)\}^3 \tag{3-30}$$

This would result in expressions for c_0 to c_3 which prove that each coefficient then depends on all calibration measurements.

Additional calibration steps can be built up in the same manner as described, increasing the order of the polynomial correction term. The equations can be generalized, and the expression for the n-th calibration function is given by:

$$h_n(x) = h_{n-1}(x) + a_n \cdot y_{ref} \cdot \prod_{i=1}^{n-1} \left(\frac{h_i(x) - y_i}{y_{ref}} \right) \tag{3-31}$$

The calibration coefficient a_n is calculated based on the n-th calibration measurement, giving $h_1(x_n)$ to $h_{n-1}(x_n)$, expressed as:

$$a_n = \frac{y_n - h_{n-1}(x_n)}{y_{ref}} \cdot \prod_{i=1}^{n-1} \left(\frac{y_{ref}}{h_i(x_n) - y_i} \right) \tag{3-32}$$

Four calibration steps are illustrated in Fig.3-13 on page 76. Clearly, the method achieves that the calibrated transfer function progresses the desired function in a *step by step* manner. The number of steps and the measurement points x_n are not necessarily fixed beforehand.

The reduction of the error signal, the difference between the calibrated output and the desired output, $\varepsilon(x)=h_n(x)-g(x)$, depends on the number of steps and the choice of the calibration points on the signal range:

1. *When taking only one calibration step, the calibration measurement is best taken with a reference input signal x_1 at the middle of the input range.*
2. *When taking two calibration steps, the calibration measurements should be done for input signals at about 25% and 75% of the input range.*
3. *For three or more calibration steps, generally a good linearization is obtained when the calibration points are selected in the following sequence: the first point at one end of the sensor range of operation, the second point at the other end of the range, and further calibration points halfway between two previously selected points.*

The method has been successfully tested on many different fictive sensor transfer functions. An example is shown in Fig.3-14, where the sensor curve was expressed as $y=f(x)=-1.2\cdot ln(1.7-x)$, and the desired sensor curve as $y=g(x)=x$. The input signal range from -1 to +1 was considered. The calibration measurements were taken respectively at $x_1=-1$, $x_2=1$, $x_3=0$, $x_4=+^1/_2$, and $x_4=-^1/_2$. The figure shows the calibrated transfer function after one step $h_1(x)$, after two steps $h_2(x)$, after three steps $h_3(x)$, after four steps $h_4(x)$, and $h_5(x)$.

The corresponding error curves $\varepsilon_n(x)=h_n(x)-g(x)=h_n(x)-x$, are shown in the pictures of Fig.3-15. The initial and the first 3 calibrated error curves, then, the third and fourth error curve, and then, the fourth and fifth error curve are displayed, using different scales. We see that after calibration of the offset and gain, a linearity error of approximately 30% (peak-to-peak band) remains for this case, which can be reduced to approximately 6% by the third calibration step, and to less than 0.5% by the fourth calibration step, and even down to 0.2% by the fifth step.

The reduction of the linearity error that can be obtained depends, of course, on the initial nonlinearity of the sensor transfer function. When the sensor transfer function has a well-known typical nonlinearity, the selection of calibration measurements can be optimized accordingly. But a sensor-characteristic linearization would then be a better approach than the

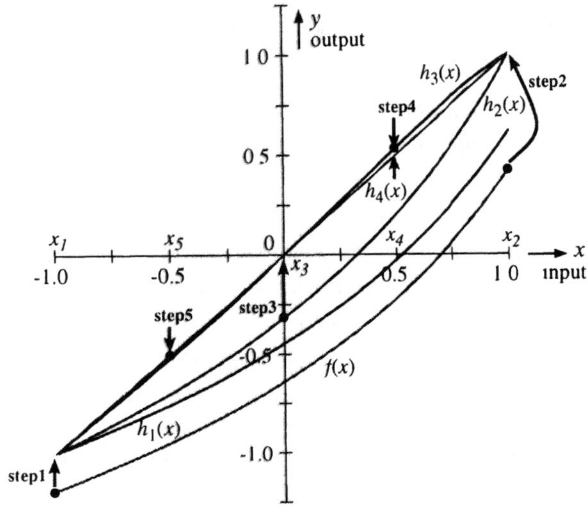

Fig. 3-14 *Example of a sensor transfer function, and the succeeding calibrated transfer functions after 1, 2, 3, 4, and 5 steps.*

polynomial method. The proposed polynomial calibration method is valuable for sensors which may show an arbitrary linearity error.

Advantages of the proposed method are:

- *a good linearization is obtained through a minimum number of calibration measurements*
- *a low number of calibration coefficients, thus only a small memory is required*
- *the required calibration coefficients can be obtained directly, without mathematical iterations*
- *a systematic step-by-step calibration, using a repetitive algorithm*

Signal Flow Diagram

The last advantage becomes very clear when the signal processing is displayed in a flow diagram as shown in Fig.3-16. The sensor output signal *y*, exhibiting a transfer function *f(x)* with respect to the physical sensor input signal *x*, is processed through a network of adders,

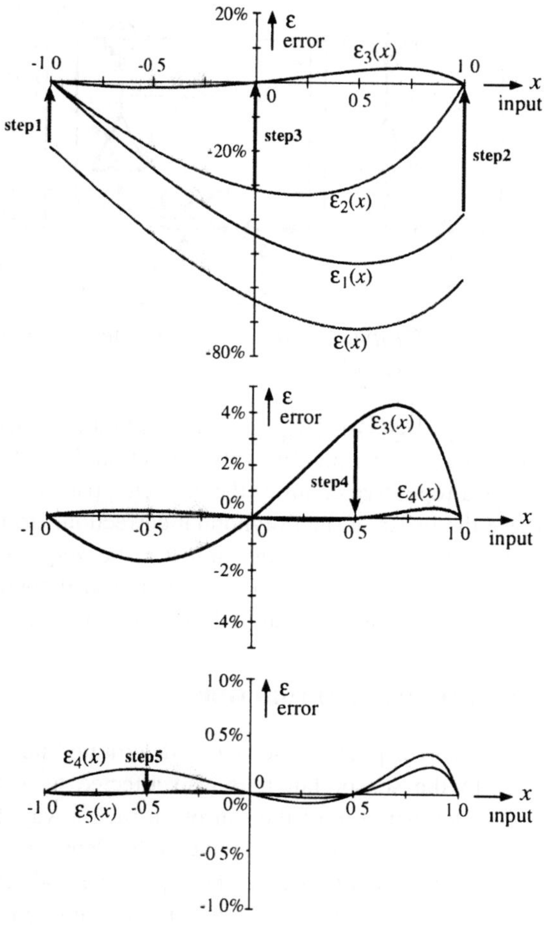

Fig. 3-15 *Error curve of the example transfer function, before and after 1, 2, 3, 4, and 5 calibration steps.*

multipliers, and trimmable calibration factors a_1 to a_n. First, all calibration factors are set to zero so that the sensor signal $y=f(x)$, left in the diagram, goes straight through to the output signal $h_4(x)$ at the right side of the diagram. Then a first calibration measurement is done by providing reference input $f(x_1)$, and adjusting a_1 to obtain an offset calibrated signal $h_4(x)=h_1(x)=f(x)+a_1 \cdot y_{ref}$, which should be equalized to the desired value y_1. A second calibration is done at the next reference input x_2 by adjusting a_2, which will add a factor $a_2 \cdot \{h_1(x)-y_1\}$ to $h_1(x)$ so that a gain-calibrated function $h_4(x)=h_2(x)$ is obtained. The third part adds the term

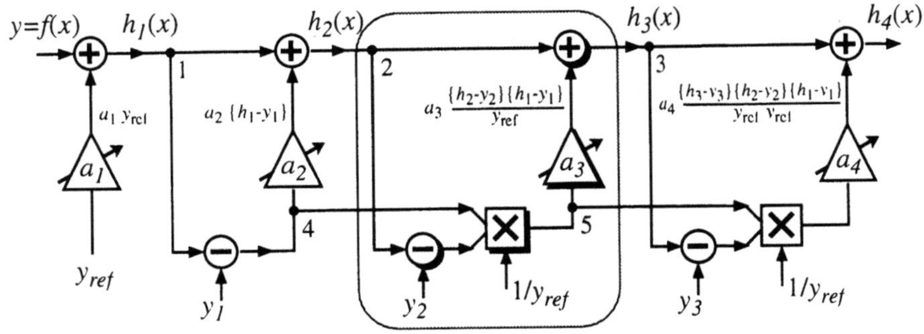

Fig. 3-16 *Graphical diagram of the polynomial calibration method.*

$a_3 \cdot \{h_1(x)-y_1\} \cdot \{h_2(x)-y_2\}/y_{ref}$ to the calibration function, to realize a second-order polynomial linearity correction. From there the flow diagram becomes a repetitive chain as the encircled part in Fig. 3-16 is repeated to increase the order of the polynomial correction. Because of the repetitive character, the calibration algorithm can be very well implemented in a (microcontroller) software routine. But it is also possible to implement the flow diagram in hardware, digital as well as analog.

3.3.2 Two-dimensional functions

The output of a pressure sensor is determined not only by the pressure applied to the sensor but it is also affected to a certain extent by the operating temperature of the sensor, in such a way that the errors, offset, gain, and nonlinearity are temperature dependent [22,23,24]. Such a sensor has to be calibrated for both pressure and temperature, hence the term two-dimensional calibration. The calibration method explained in the previous section can be extended to handle two-dimensional calibration [25,26]. The basic principle is to select lines or 'axes' in one dimension by fixing the value for one input variable, and then proceed along each selected axis with different values for the other variable, as in the one-dimensional calibration. The values for the first variable must be selected according to the same procedure as for the one-dimensional calibration.

To explain this principle mathematically, we will use normalized input variable x for the desired sensitivity, and z for the cross-sensitivity variable. The uncalibrated sensor transfer function is given by $y=f(x,z)$. To calibrate the cross-sensitivity for signal z, we assume we can make use of an additional sensor which senses z independent of other variables, $z'=k(z)$. The sensor transfer function will be calibrated by the use of:

1. *$N \times M$ calibration measurements of the sensor output at different input 'vectors', denoted by $f(x_n, z_m)$, with $n=1$ to N and $m=1$ to M*
2. *M simultaneous measurements of the output of the additional z-sensor, denoted by $z'_m=k(z_m)$, $m=1$ to M*
3. *the N desired values for the sensor output signal, according to the ideal linear transfer curve $y_n=g(x_n)$, $n=1$ to N*

After each calibration measurement, a corrected transfer curve $h_{nm}(x,z)$ is constructed, based on the preceding calibrated transfer functions and a calibration coefficient a_{nm}. Each coefficient can be calculated on the basis of the additional values obtained in the list above. Here, we will only show the calibration functions, but not the equations for calculation of the calibration coefficients.

We start at a fixed value x_1 for input variable x. We calibrate offset and the cross-sensitivity of the offset to variable z, by applying M different conditions z_1 to z_M, and calculating the calibrated transfer functions respectively as:

$$h_{11}(x, z) = f(x, z) + a_{11} \tag{3-33}$$

$$h_{12}(x, z) = h_{11}(x, z) + a_{12} \cdot (z' - z'_1)$$

$$h_{13}(x, z) = h_{12}(x, z) + a_{13} \cdot (z' - z'_1)\ (z' - z'_2)$$

$$\ldots$$

$$h_{1M}(x, z) = h_{1,M-1}(x, z) + a_{1M} \cdot \prod_{m=1}^{M-1} (z' - z'_m)$$

Each coefficient a_{1m} can be calculated by using one calibration measurement $f(x_1,z_m)$, and the desired output value for $h_{1m}(x_1,z_m)$ which is $g(x_1,z)=y_1$ in all z-points.

The set of equations in (3-33) can be compared with equation (3-22) in which the reference is left out and the coefficient a_1 is given an M-point polynomial calibration with respect to variable z.

We have now achieved that, at $x=x_1$, the calibrated transfer function $h_{1M}(x_1,z)$ equals to y_1 more or less independent of z. In the next set of calibration steps, we will rotate the transfer function around the line $(x=x_1, y=y_1)$ which runs along the z-axis, see Fig.3-4 on page 58. This way we adjust the gain of the transfer function, without disturbing the previous calibration steps. To calibrate the gain we apply a second reference for the input variable $x=x_2$. The cross-sensitivity of the gain for variable z is calibrated by applying M different conditions z_1 to z_M and computing:

$$h_{21}(x,z) = h_{1M}(x,z) + a_{21} \cdot \{h_{1M}(x,z) - y_1\}$$

$$h_{22}(x,z) = h_{21}(x,z) + a_{22} \cdot \{h_{1M}(x,z) - y_1\}(z' - z'_1)$$

$$h_{23}(x,z) = h_{22}(x,z) + a_{23} \cdot \{h_{1M}(x,z) - y_1\}(z' - z'_1)(z' - z'_2)$$

$$\ldots$$

$$h_{2M}(x,z) = h_{2,M-1}(x,z) + a_{2M} \cdot \{h_{1M}(x,z) - y_1\} \prod_{m=1}^{M-1} (z' - z'_m) \qquad (3\text{-}34)$$

The coefficients a_{2m} can be calculated by using the calibration measurements $f(x_2,z_m)$ and the desired output value in all points $g(x_2,z)=y_2$.

Again, the set of equations (3-34) can be compared with the one-dimensional gain calibration of equation (3-24) in which the calibration coefficient a_2 is given an M-point polynomial cross-sensitivity compensation for variable z.

Next, a linearity calibration is obtained by adding a second-order correction term which is zero along the two previous calibration lines

($x=x_1, y=y_1$) and ($x=x_2, y=y_2$), and by measuring at a third series of input conditions $x=x_1$ and $z=z_1$ to z_M:

$$h_{31}(x, z) = h_{2M}(x, z) + a_{31} \cdot \{h_{1M}(x, z) - y_1\}\{h_{2M}(x, z) - y_1\}$$

$$h_{32}(x, z) = h_{31}(x, z) + a_{32} \{h_{1M}(x, z) - y_1\}\{h_{2M}(x, z) - y_2\} (z' - z'_1)$$

$$h_{33}(x, z) = h_{32}(x, z) + a_{33} \cdot \{h_{1M}(x, z) - y_1\}\{h_{2M}(x, z) - y_2\} (z' - z'_1)(z' - z'_2)$$

...

$$h_{3M}(x, z) = h_{3,M-1}(x, z) + a_{3M} \cdot \{h_{1M}(x, z) - y_1\}\{h_{2M}(x, z) - y_2\} \prod_{m=1}^{M-1} (z' - z'_m)$$

(3-35)

In the same manner, the procedure can be continued to further calibrate the nonlinearity. Similar to the one-dimensional case, new input conditions for the input variable x need to be used, $x=x_4$ to $x=x_N$ in combination with the z-conditions described before. The Nth and last calibration can then be described by the following equations:

$$h_{N1}(x, z) = h_{N-1,M}(x, z) + a_{N1} \prod_{n=1}^{N-1} \{h_{nM}(x, z) - y_n\}$$

$$h_{N2}(x, z) = h_{N1}(x, z) + a_{N2} (z' - z'_1) \prod_{n=1}^{N-1} \{h_{nM}(x, z) - y_n\}$$

$$h_{N3}(x, z) = h_{N2}(x, z) + a_{N3} (z' - z'_1)(z' - z'_2) \prod_{n=1}^{N-1} \{h_{nM}(x, z) - y_n\}$$

..

$$h_{NM}(x, z) = h_{N,M-1}(x, z) + a_{NM} \prod_{m=1}^{M-1} (z' - z'_m) \prod_{n=1}^{N-1} \{h_{nM}(x, z) - y_n\}$$

(3-36)

After all the coefficients a_{11} to a_{NM} have been calculated, the final calibration function, $h(x,z)=h_{NM}(x,z)$, has been given an $(N\text{-}1)^{th}$-order polynomial linearization for input variable x, and an $(M\text{-}1)^{th}$-order compensation for cross-sensitivity to variable z. If the linearization has to be improved, the polynomial order for input x can be increased with an additional set of (M) equations and measurements as expressed in equation (3-36). Only the new calibration coefficients need to be calculated, the previous ones remain valid. On the other hand, if the order of the cross-sensitivity polynomial needs to be increased, then each set of equations (3-33) to (3-36) gets one additional formula. We need N additional calibration measurements for inputs x_1 to x_N at a new condition z_{M+1}. It can be shown that, despite the nested (recursive) use of the functions $h_{nm}(x,z)$, the additional calibration coefficients can be calculated without having to recompute the previous calibration functions.

Practical implications and results of the two-dimensional calibration will be presented in chapter 6.

3.3.3 Multi-dimensional functions

In the same way as the polynomial calibration has been extended from one to two dimensions, it can be extended to a third dimension. All the calibration coefficients a_{nm} in equations (3-33) to (3-36) then need to be calibrated for sensitivity to the third variable in a 'nested' series of measurements. It is also possible to put 2 two-dimensional calibrations 'in series', first calibrating for one cross-sensitivity as described above, then calibrate the resulting transfer curve for the other cross-sensitivity. Besides the mathematical complications, the practical complication of the $N \times M \times L$ calibration measurements with three controlled physical input signals will make such a three-dimensional calibration very unattractive. Using vector and matrix expressions it is possible to expand the polynomial calibration to even more dimensions, but it would be more expedient to concentrate on elementary ways to reduce the sensor's cross-sensitivity to so many parameters.

3.4 Conclusion

The discrepancies that can be found in the sensor transfer have been distinguished to clarify what errors can be corrected by means of calibration. Generally, the aim of the calibration is to correct the sensor transfer for linear errors, like offset and gain (scale), as well as nonlinearity and cross-sensitivity.

Several linearization methods have been explained and applied to an sensor transfer function example. Without prior knowledge about the sensor characteristic, a general curve-fitting method offers the possibility to linearize the sensor transfer, using a low number of calibration measurements. The polynomial curve-fitting technique has been adapted in such a way that the sensor curve can be corrected right after each calibration step, without affecting previous calibration steps. The corrected sensor transfer function then progresses towards the desired (linear) transfer in a step-by-step manner. The proposed method can be extended to two-dimensional transfer functions, which enables a systematic calibration of linearity and cross-sensitivity.

References

[1] J.W. Gardner, *Microsensors: principles and applications*, John Wiley & Sons, New York, 1994.

[2] W. Gopel, J. Hesse, and J. Zemel, "*Sensors: A Comprehensive Study, Vol.1: Fundamentals and General Aspects*", T. Grandke & W.H. Ko (Vol.eds.), *VCH*, Weinheim, 1989.

[3] J.H. Huijsing and J.A. van Steenwijk, "A monolithic analog exponential converter", *IEEE Journal of Solid-State Circuits*, Vol.15, No.2, 1980, pp.162-168.

[4] J.E. Brignell, "Digital compensation of sensors", *Journal of Physics*, Vol. E: Scientific Instrumentation, 1987, pp.1097-1102.

[5] J.E. Brignell, "Software techniques for sensor compensation", *Sensors and Actuators A*, Vol.25-27, 1991, pp.29-35.

[6] P.P.L. Regtien and P.J. Trimp, "Dynamic calibration of sensors using EEPROMs", *Sensors and Actuators A*, Vol.21-23, 1990, pp.615-618.

[7] P.N. Mahana and F.N. Trofimenkoff, "Transducer output signal processing using an eight-bit microcomputer", *IEEE Trans. Instrumentation and Measurement*, Vol.IM-35, No.2, June 1986, pp.182-186.

[8] C. de Boor, "A practical guide to splines", *Applied Mathematical Sciences*, Vol.27, Springer, New York, 1978.

[9] P. Dierckx, *Curve and surface fitting with splines*, Oxford University Press, Oxford, 1993.

[10] S.B. Crary, W.G. Baer, J.C. Cowles, and K.D. Wise, "Digital compensation of high performance silicon pressure transducers", *Sensors and Actuators A*, Vol.21-23, 1990, pp.70-72.

[11] S. Huang, R.Z. Morawski, and A. Barwicz, "Static calibration based on superposition of splines in one variable", *Proceedings IMTC'96*, June 1996, pp.49-53.

[12] C. Berthoud, M. Ansorge, and F. Pellandini, "Effective static response compensation suitable for low-power asic implementation with an application to pressure sensors", *Proceedings IMTC'96*, June 1996, pp.1168-1173.

[13] M.E. Snow and S.B. Crary, "The use of simulated annealing in the I-optimal design of experiments", *Michigan Academician XXIV*, 1992, pp.343-354.

[14] S.B. Crary, L. Hoo, and M. Tennenhouse, "I-Optimality algorithm and implementation", *Computational Statistics*, Y. Dodge and J. Whittaker (eds.), Proceedings of the 10th Symposium on Computational Statistics, Vol.2, Neuchâtel, Switzerland, 1992, pp.209-213.

[15] S.R. Ashley, M. Muggeridge, and J. Lucas, "An inexpensive digital linearizer for nonlinear transducers", *Journal of Physics E: Sci. Instrum.*, Vol.11, 1978, pp.576-580.

[16] W.T. Bolk, "A general digital linearising method for transducers", *Journal of Physics*, Vol. E: Scientific Instrumentation, 1985, pp.61-64.

[17] D. Patranabis and D. Gosh, "A novel software-based transducer linearizer", *IEEE Trans. Instrumentation and Measurement*, Vol.36, No.6, December 1989.

[18] P. Malcovati, C.A. Leme, P. O'Leary, F. Maloberti, and H. Baltes, "Smart sensor interface with A/D conversion and programmable calibration", *IEEE Journal of Solid-State Circuits*, Vol.29, No.8, August 1994, pp.963-966.

[19] M. Yamada and K. Watanabe, "A capacitive pressure sensor interface using oversampling Δ–Σ demodulation techniques", *IEEE Trans. Instrumentation and Measurement*, Vol.46, No.1, February 1997, pp.3-7.

[20] M. Gunawan, G.C.M. Meijer, J. Fonderie, and J.H. Huijsing, "A curvature-corrected low-voltage bandgap reference", *IEEE Journal of Solid-State Circuits*, Vol.28, No.6, 1993, pp.667-670.

[21] S. Kaliyugavaradan, P. Sankaran, and V.G.K. Murti, "A new compensation scheme for thermistors and its implementation for response linearization over a wide temperature range", *IEEE Trans. Instrumentation and Measurement*, Vol.42, No.5, 1993, pp.952-956.

[22] S. Kim and K.D. Wise, "Temperature sensitivity in silicon piezoresistive pressure transducers", *IEEE Trans. Electron Devices*, Vol. ED-38, 1983, pp.802-810.

[23] H.-J. Kress, F. Bantien, J. Marek, and M. Willmann, "Silicon pressure sensor with integrated CMOS signal-conditioning circuit and compensation of temperature coefficient", *Sensors and Actuators A*, Vol.25-27, 1991, pp.21-26.

[24] M. Akbar and M.A. Shanblatt, "Temperature compensation of piezoresistive pressure sensors", *Sensors and Actuators A*, Vol.33, 1992, pp.155-162.

[25] K.F. Lyahou, G. v.d. Horn, and J.H. Huijsing, "A non-iterative polynomial 2-dimensional calibration method implemented in a microcontroller", *Proceedings IMTC'96*, June 1996, pp.62-67.

[26] G. v.d. Horn and J.H. Huijsing, "Integrated smart sensor calibration", *J. Integrated Analog Circuits and Signal Processing*, Vol.14, No.3, November 1997.

Calibration using Analog Signal Processing

4

4.1 Introduction

In this chapter we will focus on calibration techniques based on analog signal processing. Almost all sensors require some kind of analog signal conditioning or amplification before the sensor signal can be converted and processed digitally. Basically, sensors produce an analog output signal only. But even for modern smart sensors with digital output, a standardized analog output signal is often desired as well. Therefore, we will first consider the possibilities of adjusting the transfer of an analog circuit for the purpose of calibration.

Adjustable elements, like potentiometers or laser-trimmed resistors, can be used. An example will be explained in section 4.2. As mentioned in previous chapters, it is preferable to realize a digitally programmable calibration. This would allow *batch calibration* and *in-situ calibration*, and thus simplify *automation* of the (factory) calibration. It would also enable calibration of the sensors after packaging and allow (customer) *re-calibration*, and thus improve sensor performance. The laser-trimmed resistors in the conventional calibration circuit could be replaced by resistor arrays with switches, but instead we prefer to make use of Digital-to-Analog converters, as will be explained in section 4.3. Pulse-width and pulse-density modulation are very promising techniques for calibration, as will be shown in section 4.4. The examples in those sections will be limited to calibration for offset, gain, and the linear temperature coefficients of offset and gain. In section 4.5, it will be demonstrated that the polynomial linearization explained in the previous chapter can also be implemented in a programmable analog circuit.

4.2 Conventional sensor calibration

Before calibrating the transfer of a sensor electronically, other options to improve the sensor characteristics should be considered first; the less the size and number of errors to be calibrated the better the final performance will be. If cross-sensitivity to another physical parameter can be eliminated systematically in the sensor construction, this is preferable to cross-sensitivity compensation that requires calibration. Nevertheless some errors will always remain, and require calibration. Often the sensor output can only be related to the physical input signal through a calibration. The transfer may then be corrected by processing the electrical output signal, or for some sensors (modulating sensors), by modifying the electrical reference signal.

4.2.1 Resistor trimming

In discrete electronics, most passive elements are also available in mechanically adjustable implementations: potentiometers, variable capacitors, etc. An advantage of such elements is that they have inherent analog memory: potentiometers stay in their last position. Comparable adjustments in Integrated Circuit (IC) technology make use of laser trimming. A laser beam can be used to modify the geometrical properties of a resistor on the chip surface, in order to adjust the resistance value. An obvious disadvantage of the laser-trimming technique is that it alters the resistors permanently.

For example, with sensors based on a resistive bridge configuration it makes sense to use trimmable resistors. Fig.4-1 shows a Wheatstone bridge in which some extra resistors, R_1 to R_6, have been inserted [1,2,3]. The resistor combination R_1, R_2 makes it possible to adjust the bridge balance to calibrate offset. Resistor R_3 is a temperature-dependent resistor

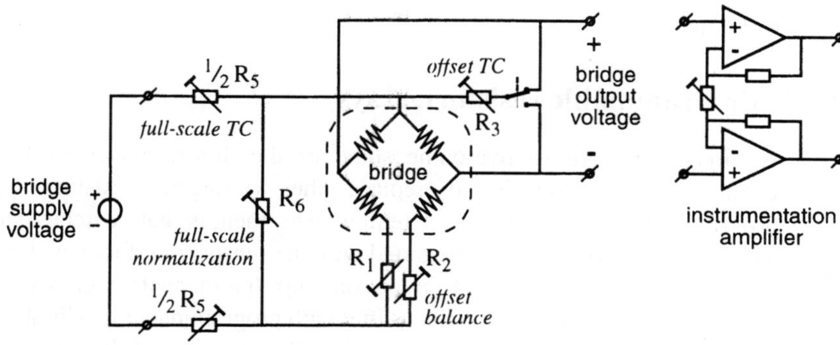

Fig. 4-1 *Wheatstone bridge with resistor compensation.*

which may be adjusted to compensate for the offset temperature coefficient. The voltage divider R_5, R_6 realizes an adjustment of the bridge reference voltage and its temperature gradient, so that gain errors can be compensated and the full-scale output (FSO) can be normalized. The voltage divider can only reduce the bridge voltage and thus decrease the signal. With normalization of the FSO, all sensors will thus be adapted to the one with the lowest sensitivity. It is preferable to obtain a gain-adjustment without reducing the bridge reference. This could be done with a laser-trimmed feedback resistor in an instrumentation amplifier as shown in Fig.4-1.

Disadvantages of laser-trimming calibration techniques are:

- *the adjustments are not independent of each other (the offset resistors also affect the gain)*
- *resistors can often only be changed one way, usually resistance is increased*
- *the trimming is irreversible, so no re-calibration is possible*
- *in-situ calibration (under test) is difficult or impossible*
- *automation in a fabrication process is difficult*
- *it is a time-consuming expert operation, thus expensive*

An advantage of laser-trimming, and also a reason why it is still common practice, is that, besides the resistors, no other elements need to be integrated with the sensor. If enough research is done to make IC technology and sensor technology compatible, the integration of active elements with the sensor may not be a problem in the future. Smart sensor electronics will then provide more attractive solutions than laser-trimming techniques.

4.2.2 Programmable resistor-arrays

A direct approach to overcome some of the disadvantages of laser-trimming calibration is to replace the trimmable resistors with programmable resistor arrays. A requirement, then, is that switches, such as MOS transistors, can be integrated with the sensor (interface), and their position/state can be stored in an on-chip (permanent) memory. An example of an instrumentation amplifier with programmable gain is shown in Fig.4-2. The feedback is built-up with an array of binary-scaled

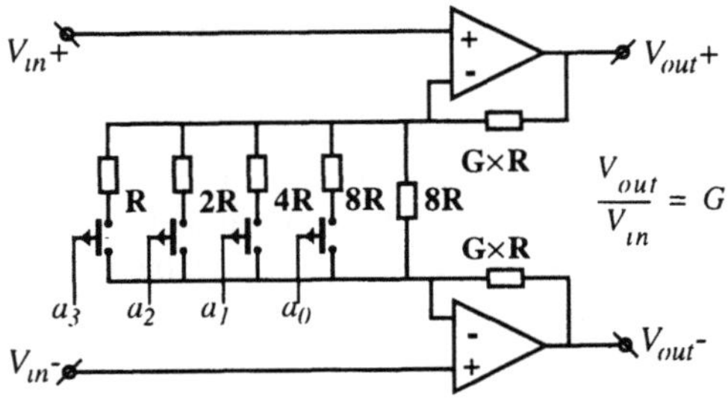

$$\frac{V_{out}}{V_{in}} = G\left(\frac{1}{16} + a_0 \cdot \frac{1}{16} + a_1 \cdot \frac{1}{8} + a_2 \cdot \frac{1}{4} + a_3 \cdot \frac{1}{2}\right)$$

Fig. 4-2 *Instrumentation amplifier with programmable feedback resistor.*

resistors that can be switched in parallel. The result is that the voltage gain of the amplifier can be programmed digitally, in this case with a 4-bit resolution. Likewise, other trimmable resistors in conventional configurations could be replaced with programmable resistors. However, if IC-technology is available, it is better to investigate more advanced circuit techniques rather than to adapt conventional sensor calibration techniques and inherit many of their disadvantages.

4.3 Analog calibration circuits

When an analog signal has to be influenced in a programmable way the use of some sort of Digital-to-Analog (DA) conversion is unavoidable[1]. Despite the many DA-conversion techniques that exist and are being developed [4], we will concentrate on the use of a conventional multiplying DA-converter (DAC) configuration.

The principle of this DA-conversion is based on the creation of binary-scaled portions of an analog reference signal, which can then individually be switched on or off by the bits of the digital input. The parts that are switched on need to be summed together in the analog signal domain somehow. Current signals can be summed easily on a common low-impedance node, whereas voltage signals have to be made time-discrete (sampled) before they can be summed in a switched-capacitor circuit. When sampling of the analog signal has to be avoided, it is logical to create binary-scaled currents and use current switches. Fig.4-3 shows a simple 4-bit example of a multiplying DAC in both CMOS implementation and bipolar implementation. Instead of connecting an analog reference, we apply an input signal. The operation will be explained in short, so that the application of the DACs in the next section will be clear.

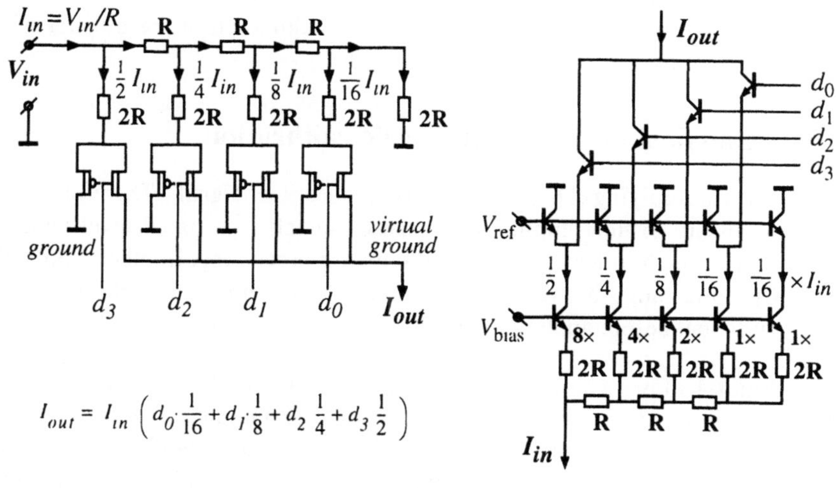

Fig. 4-3 *Simple 4-bit multiplying DAC examples, using CMOS voltage switches (left), or bipolar current switches (right).*

[1] unless reliable analog electrical memories become available

The CMOS circuit on the left uses a resistor ladder network to obtain binary-scaled currents. Depending upon the digital bit values, the vertical resistors are either short-circuited to ground by the PMOS transistor (bit *low*) or to a virtual ground node by the NMOS transistor (bit *high*). The effective resistance of the ladder network from input to ground equals R. The input voltage signal results in an input current $I_{IN}=V_{IN}/R$ which scales down by a factor of two at each successive node in the ladder. The input of an IV-converter can be used as a virtual ground and summing node for the binary parts.

In the bipolar version, a cascode with scaled transistors is used to create the binary portions of the current. If resistor matching is better than transistor matching, the emitter resistor network as shown will improve the accuracy of the binary division. The currents are switched by differential pairs which are controlled by digital signals with limited amplitude.

The formula in Fig.4-3 shows that the analog input current is multiplied by the digital number applied at the switches. The next section will show how such multiplying DACs can be used for calibration of the transfer of an analog sensor read-out circuit.

4.3.1 Classical offset and full-scale calibration

The multiplying DACs can only reduce the signal. Therefore, it is necessary to first amplify small sensor signals to a signal level which is high enough to leave some 'headroom' for processing. In other words, the dynamic range between signal amplitude and noise+distortion must be sufficient before applying the signal to the multiplying DACs.

Fig.4-4 shows a straightforward application of the DACs to make the gain and offset of an analog read-out circuit programmable. The sensor output signal V_{sensor}, presumably a voltage[1], is amplified or buffered at the input of the circuit and the voltage signal is then converted by the DAC into a current. The magnitude of the current is controlled by the switches of the DAC. The position of the switches can be stored in a very small and simple digital memory (register), which is not shown in the figure. The memory should be programmable and preferably non-volatile.

[1] Other kind of sensor signals could be converted to a voltage signal, instead of just amplified

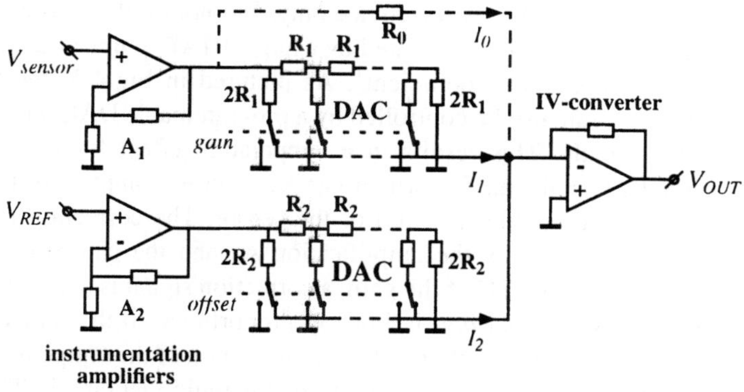

Fig. 4-4 *Analog calibration circuit, using multiplying DACs for a programmable gain and offset control.*

The output current of the DAC is converted back into a voltage by the IV-converter. The digital number applied at the switches of the DAC controls the gain of the transfer. When there are no other signal paths, the gain can be varied over the full range, from zero to the maximum. Often the gain needs only a small adjustment relative to the full range. This can be achieved by connecting a resistor R_0 across the DAC, as illustrated with the dashed wires. Now an additional current I_0 proportional to the sensor signal will contribute to the output voltage. Depending on the ratio of R_1 and R_0, only a fraction of the full-scale signal (gain) can be varied.

Correction of offset of the transfer requires a reference which is independent of the sensor signal. The reference voltage V_{REF} is treated in the same manner as the input voltage which results in a programmable current I_2. The current is added to the other signals at the low-impedance input node of the IV-converter, and thus realizes a shift of the output voltage level. The offset correction range depends upon the value of the reference voltage and the ratio of R_2 to R_1. Here, too, a resistor could be placed across the DAC in order to create an additional prefixed offset correction (not shown in figure).

4.3.2 Classical cross-sensitivity calibration

Clearly the circuit construction of Fig.4-4 allows the addition of many correction signals at the input of the IV-converter. For example, this can be done to realize a correction for cross-sensitivity to temperature, provided that a signal is available which depends linearly on the operating temperature of the sensor device only, and not on the sensor signal itself. Such a signal can be generated by using a PTAT source or a resistor with a linear temperature coefficient[1]. As pictured in Fig.4-5, the temperature signal V_T can also be controlled by a programmable DAC and added to the other signals. This results in a temperature effect for the offset of the circuit's signal transfer, which can be used to compensate for the offset temperature coefficient (TC) of the sensor. The correction range for the offset TC is set by the amplification A_4 and the choice of R_4. For the correction of the TC of the gain, a correction signal is used which depends on the input signal *and* is linearly proportional to the temperature. This correction signal is obtained by multiplication of the temperature signal V_T and the sensor signal V_{sensor} in an analog multiplier circuit. The multiplier output voltage signal is also converted to a current, controlled by a DAC, and added to the other signals.

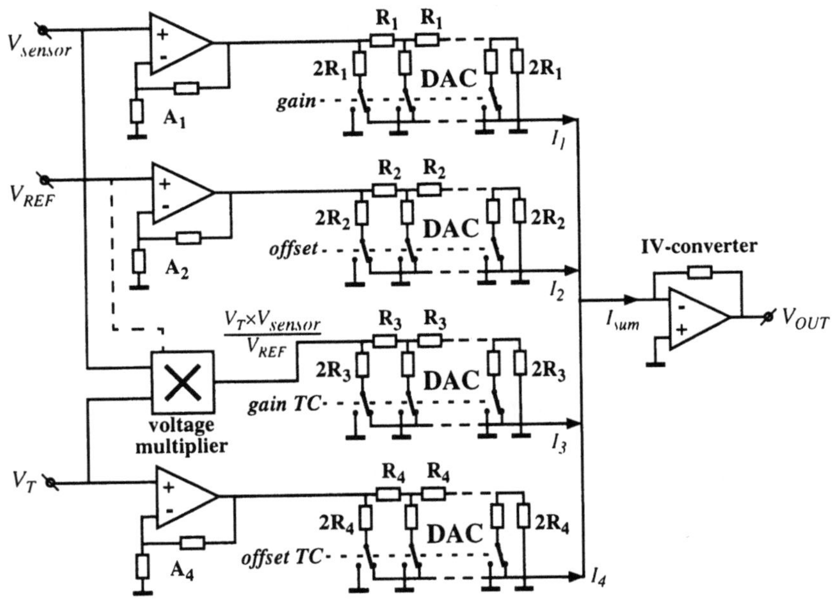

Fig. 4-5 *Analog circuit with multiplying DACs for offset, gain, gain TC, and offset TC control.*

[1] For piezoresistive bridges with a constant current excitation, sometimes the bridge voltage can be measured and used as a temperature signal

In principle, the four DACs control the four different errors, offset, offset TC, gain, and gain TC independently. But the different correction signals may also have transfer errors which need to be measured. Therefore, in the calibration procedure, each of the four signal contributions has to be measured separately at two different physical input signals and at two temperatures (2×2 conditions). This can be done by respectively setting each DAC on maximum and the other three on zero. These 2×2 ×4 output measurements can be used to characterize all four signal contributions. Then the required values for the DACs can be calculated and programmed in the DACs, so that the desired calibration is obtained.

The schematic of Fig. 4-5 can be worked out in a bipolar implementation. In that case, we prefer to make use of *current* multiplying DACs, see the beginning of section 4.3. The input amplifiers must then be replaced by VI-converters, the current signal of a PTAT source can be used as temperature signal, and a (Gilbert) multiplier with current output can be applied for multiplication. The idea is illustrated in Fig. 4-6.

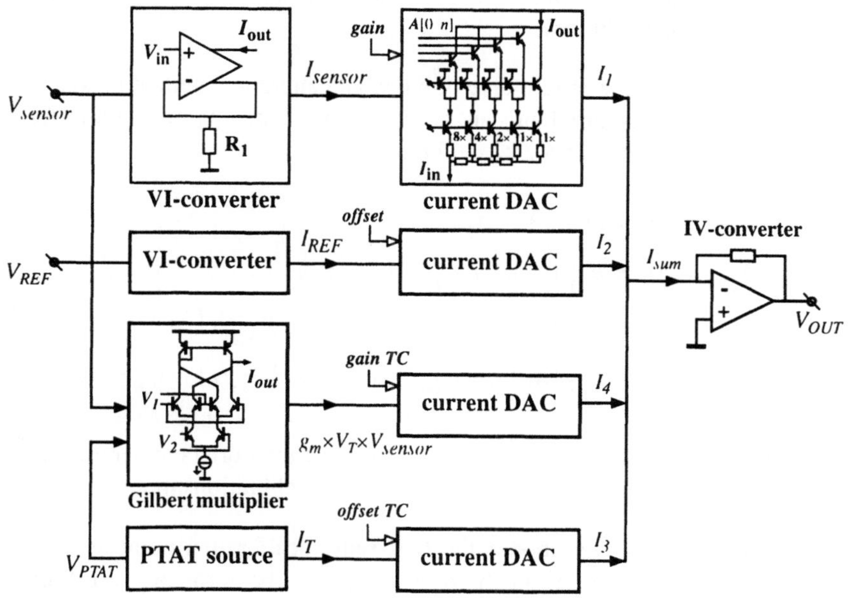

Fig. 4-6 *Bipolar implementation using current-mode DACs.*

In the schematics of Fig. 4-4 to Fig. 4-6 we have ignored the fact that the errors may have a different polarity than expected. For example, the TC of the offset can be either positive or negative. By realizing the calibration circuit in a *differential* implementation, it becomes possible to give the correction signals a positive or negative sign. The input amplifiers are replaced by differential instrumentation amplifiers, and the output IV-converter is replaced by a differential equivalent, see also section 4.5. The programmable output signals of the DACs can then be switched to either the positive or the negative input of the IV-converter to control the polarity of the correction. A differential circuit also offers an improved performance with respect to the common-mode and supply rejection ratios.

4.3.3 Improved analog calibration circuit

In the analog calibration circuits proposed in the previous section, the calibration settings are not independent of each other. This means that the offset has to be re-adjusted after a gain calibration, and the other way round. Unless the individual signal contributions are measured separately, and the required calibration factors are calculated externally, it turns into an *iterative* calibration.

According to the progressive polynomial calibration method introduced in *chapter 3*, this problem can be overcome. Using the equations for the transfer functions in a 2×2 calibration:

$$\begin{aligned}
h_{11}(x, z) &= f(x, z) + a_{11} \\
h_{12}(x, z) &= h_{11}(x, z) + a_{12} \cdot (z' - z'_1) \\
h_{21}(x, z) &= h_{12}(x, z) + a_{21} \cdot \{h_{12}(x, z) - y_1\} \\
h_{22}(x, z) &= h_{21}(x, z) + a_{22} \cdot \{h_{12}(x, z) - y_1\} \cdot (z' - z'_1)
\end{aligned} \qquad (4\text{-}1)$$

with x as the sensor input signal, and z' the measured temperature signal, we can construct the desired transfer of the calibration circuit.

For simplicity, we choose the first sensor calibration point at the input level S_1 where we want the output signal to be zero, so y_1=0. Suppose, furthermore, that we can make use of a voltage signal $V_{T\text{-}T1}$ which is zero at a known temperature T_1. Next, the functions in equation (4-1) are replaced by voltage signals. Then, we can derive that the following transfer is desired:

$$V_{12} = V_{sensor} + a_{11} \cdot V_{REF} + a_{12} \cdot V_{T-T0}$$

$$V_{OUT} = V_{22} = V_{12} + a_{21} \cdot V_{12} + a_{22} \cdot \frac{V_{12} \times V_{T-T0}}{V_{REF}} \quad (4\text{-}2)$$

This transfer can be implemented in an analog calibration circuit as illustrated in Fig. 4-7. It is, in fact, a slight modification of the original circuit shown in Fig. 4-5.

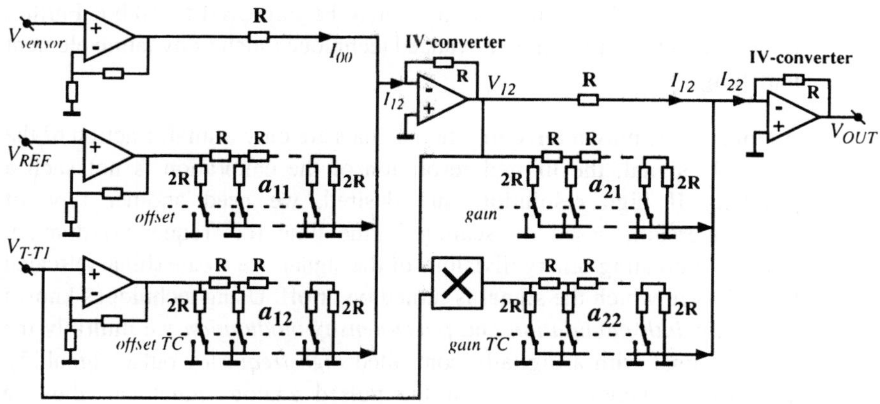

Fig. 4-7 *Alternative analog circuit for independent calibration of offset, gain, and temperature coefficients.*

The calibration can now be executed in the manner of the progressive calibration method. First, all the 'DAC coefficients' are set to zero. At temperature T_1, and a zero physical input signal S_1 for the sensor, DAC coefficient a_{11} is adjusted in such a way that the desired zero output voltage is obtained. Second, at another temperature T_2, a_{12} is adjusted to compensate the temperature gradient in the offset of the sensor. Now the signals I_{12} and V_{12} are zero for a zero sensor input signal. At the next calibration step another sensor input S_2 is provided, and at temperature T_1 the gain can be adjusted with a_{21}. At the other temperature T_2, the temperature coefficient (TC) of the gain can be compensated with a_{22}. In that way, the calibration settings can be controlled separately. This is achieved because, independent of the DAC settings, the signal contributions for the gain corrections are zero when the sensor input signal is equal to the calibration point S_1, and the signal contributions for TC compensation are zero at temperature T_1.

4.4 Classical pulse-modulated calibration

The performance of the multiplying DACs depends on the matching of the transistors and resistors used. In particular, it is the monotonicity of output versus increasing digital code that is affected by mismatches [4]. Without taking special measures the highest resolution for which monotonicity can be guaranteed is about 7-8 bits. With special transistor and resistor layouts to improve matching, the resolution may be improved to 10 bit. Further improvements require more advanced techniques such as dynamic element matching.

As long as the maximum correction signals are only a small fraction of the full-scale signal, the limited resolution of the calibration is not such a problem. If high resolutions are desired, we need another type of DA-conversion, based on scaling in the time (or frequency) domain. Instead of creating binary divisions of the signal, we create (binary) scaled time slots in which the signal is turned on or off. Using techniques known as *pulse-width modulation* and *pulse-density modulation* we multiply the analog signal with a digitally controlled *high-frequency* pulse signal [5]. The low-frequency *average* of the pulsed analog signal can thus be accurately controlled by the digital signal. The resolution is now determined by the time accuracy of the pulses and the maximum pulse width. These properties relate respectively to the stability of the digital clock signal (jitter) and the required analog signal bandwidth.

The analog calibration circuit of Fig.4-5 can be very much simplified by replacing each multiplying DAC by a single resistor or VI-converter and a (current) switch. This is displayed in Fig.4-8. Each switch is controlled separately by a digital *pulse train* or *stream.* The four pulse trains regulate the average contribution of the current signals I_1 to I_4 to the input of the IV-converter. The averaging function can be realized by placing a capacitor across the feedback resistor of the IV-converter. The low-pass filtering achieved suppresses the high-frequency contents of the pulse trains. Although the analog circuit is simplified, the complexity is now moved to the digital circuit which has to generate the four pulse train signals. There are several options for the implementation of a digital pulse

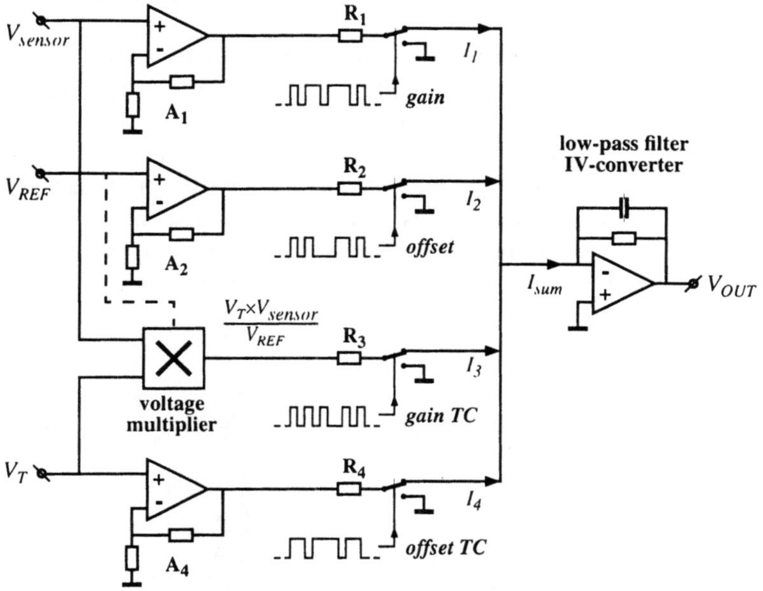

Fig. 4-8 *Analog calibration circuit using switched-current DACs.*

generator which creates a digital *1*-bit pulse signal of an *n*-bit digital number. One option is to use the digital equivalent of a sigma-delta converter, which outputs a so-called bitstream signal. This technique will be elaborated in *chapter 5* and combined with a sigma-delta AD-converter in place of the IV-converter.

The circuit of Fig.4-8 should preferably be implemented in a differential version for the same reasons mentioned in the previous section. But also to improve resistance against the digital interference, which is expected to increase considerably.

Furthermore, it must be noted that there is no longer a time-continuous analog signal path; the analog signals are interrupted by the switches. To prevent interference and aliasing effects, the bandwidth of the analog signals must be limited in such a way that it does not overlap the frequency band of the pulse-train signals.

4.5 Analog implementation of a polynomial calibration

The circuit examples discussed so far have shown the possibilities of correcting *linear* errors in the transfer of an analog circuit in a very straightforward way [6]. As discussed in *chapter 3*, there are several methods to correct the *nonlinearity* of the sensor transfer. A very effective linearization could be obtained with the presented progressive polynomial calibration method. In this section we will describe a complete analog implementation and realization of a four-step linearity calibration. For clarity, the signal flow diagram explained in section 3.3 is repeated below in Fig.4-9.

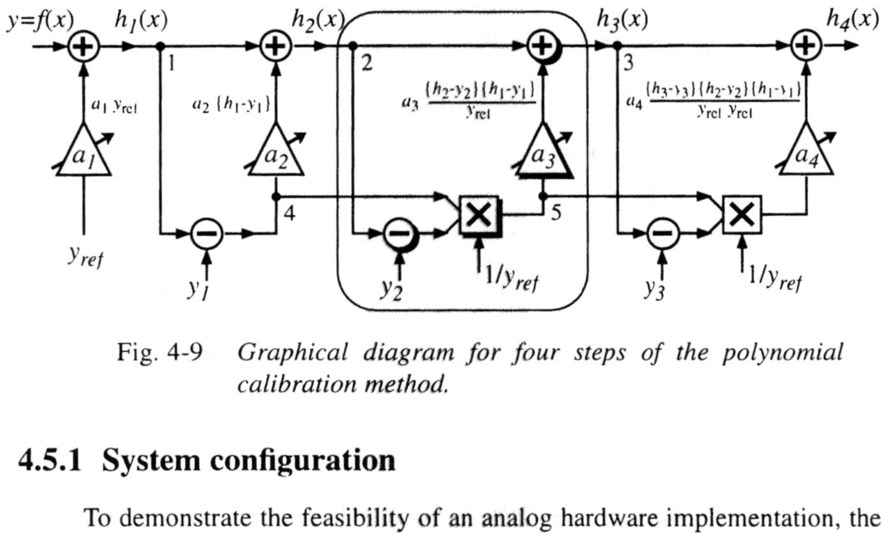

Fig. 4-9 *Graphical diagram for four steps of the polynomial calibration method.*

4.5.1 System configuration

To demonstrate the feasibility of an analog hardware implementation, the flow diagram of Fig.4-9 has been implemented as a programmable analog circuit in a low-cost bipolar process [7,8]. Fig.4-10 shows the block diagram of the analog part. When compared carefully, it shows great similarity with the diagram of Fig.4-9. All signals are represented by differential currents which are superimposed on common-mode bias currents. Current signals $i_{\mathrm{n}}(V_{\mathrm{IN}})$ in Fig.4-10 correspond to calibration functions $h_{\mathrm{n}}(x)$ in Fig.4-9; currents I_n correspond to calibration points y_{n}, and current I_{REF} to y_{ref}.

The input voltage V_{IN}, coming from the sensor output, the reference signal V_{OS}, and the set-points for the calibration V_1, V_2, V_3 are all converted to differential current signals, using Voltage-to-Current (VI) converters. At

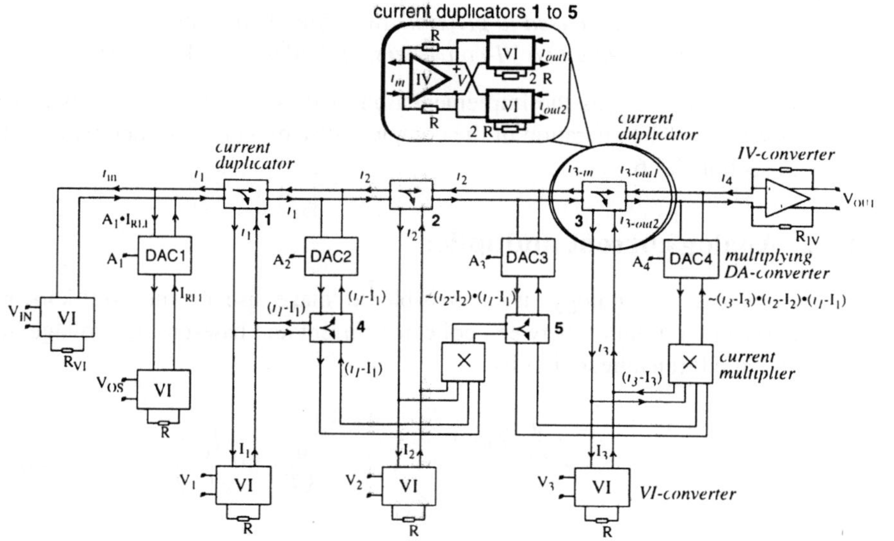

Fig. 4-10 *Programmable analog signal processor, implementation of* Fig. 4-9.

the output, a Current-to-Voltage (IV) converter turns the processed current back into a differential voltage signal. The signal processing is implemented in current-mode circuitry. The required processing functions are listed below.

- *The adding and subtracting is done by correctly connecting the differential current signals (Kirchoff's law).*
- *The multiplication of the analog current signals is implemented in a modified Gilbert multiplier, a well-known translinear circuit, using only a few transistors [9].*
- *The calibration coefficients a_1 to a_4 are implemented as multiplying Digital-to-Analog Converters (DAC), according to the principle explained in section 4.2.2. The DACs are combined with programmable digital registers which are connected to the digital inputs of the DACs, indicated by A_1 to A_4. The digital registers are not shown in* Fig. 4-10.
- *As can be seen in the flow diagram of* Fig. 4-9, *a duplication of the signal is necessary at the nodes numbered 1 to 5. This is implemented in current duplicator blocks. Inside each of*

these blocks an IV-converter turns the current into a voltage which is then used to generate two differential current signals, using two VI-converters, as explained in Fig.4-10.

In the next sections the implementation of these function blocks will be elaborated. The operation of the complete system will be further explained in section 4.5.6.

4.5.2 Analog current multiplier

In bipolar technology, it is possible to make use of the well-known exponential relation between collector current and base-emitter voltage in so-called translinear circuits:

$$I_C = I_S \cdot e^{\left(\frac{q}{kT} V_{BE}\right)} \Leftrightarrow V_{BE} = \frac{q}{kT} \cdot \ln\left(\frac{I_C}{I_S}\right) \qquad (4\text{-}3)$$

A special class of circuits is based on the principle that when the *sum of base-emitter voltages* of a set of transistors is constant, this also applies for the *product of the collector currents* [9,10]. An advantage of this principle is that it is valid for a very large range of values, because the exponential relation is a physical property of the bipolar transistor. The best known example of a translinear circuit is probably the current multiplier.

A variant of the translinear multiplier for the multiplication of two differential current signals is shown in Fig.4-11. It consists of two transistor mazes, one with Q1,Q2,Q3,Q4 and another with Q1,Q2,Q5,Q6. The transistors are connected in such a way that the sum of base-emitter voltages on both sides of a maze are equal, thus $V_{be3}+V_{be2}$ equals $V_{be1}+V_{be4}$ and $V_{be5}+V_{be2}$ equals $V_{be1}+V_{be6}$. Using equation (4-3), it results in the following translinear relation:

$$V_{be1} - V_{be2} = V_{be3} - V_{be4} = V_{be5} - V_{be6} \Leftrightarrow \frac{I_{c1}}{I_{c2}} = \frac{I_{c3}}{I_{c4}} = \frac{I_{c5}}{I_{c6}} \qquad (4\text{-}4)$$

At all four input terminals at the bottom of the circuit, currents are pulled down to the negative supply or ground. The bias voltage at the base of Q1 and Q2 determines the voltage level at the current inputs.

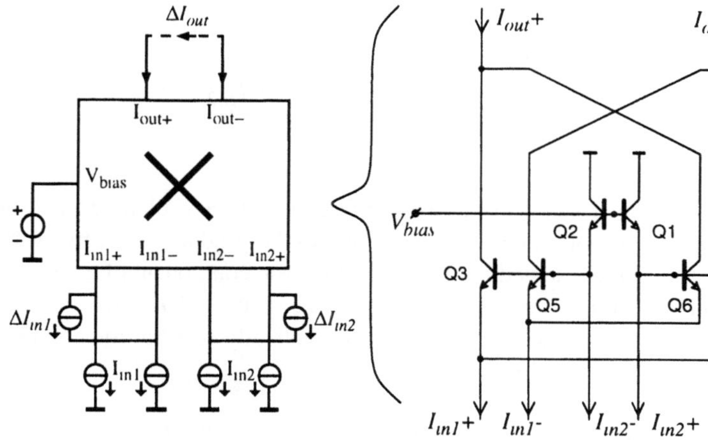

Fig. 4-11 *Four-quadrant current-mode multiplier.*

Because of the differential set-up, each input current consists of two parts, I_{in+} and I_{in-}. Each part is made up of a common-mode component, I_{in}, and a differential component, ΔI_{in}. When neglecting the base currents, we see for *input1* that $I_{in1+}=I_{e3}+I_{e4}\approx I_{c3}+I_{c4}$ and $I_{in1-}=I_{e5}+I_{e6}\approx I_{c5}+I_{c6}$ and for *input2* that $I_{in2+}=I_{in2}+\Delta I_{in2}\approx I_{c1}$ and $I_{in2-}=I_{in2}-\Delta I_{in2}\approx I_{c2}$, which can be worked out in combination with equation (4-4) into:

$$I_{c3}=\frac{1}{2}\cdot\left(1+\frac{\Delta I_{in2}}{I_{in2}}\right)\cdot I_{in1+} \qquad I_{c4}=\frac{1}{2}\cdot\left(1-\frac{\Delta I_{in2}}{I_{in2}}\right)\cdot I_{in1+}$$

$$I_{c6}=\frac{1}{2}\cdot\left(1-\frac{\Delta I_{in2}}{I_{in2}}\right)\cdot I_{in1-} \qquad I_{c5}=\frac{1}{2}\cdot\left(1+\frac{\Delta I_{in2}}{I_{in2}}\right)\cdot I_{in1-} \tag{4-5}$$

Furthermore, we see in Fig.4-11, for the two parts of the output current that $I_{out+}=I_{c3}+I_{c6}$ and $I_{out-}=I_{c4}+I_{c5}$. When substituting $I_{in1+}=I_{in1}+\Delta I_{in1}$ and $I_{in1-}=I_{in1}-\Delta I_{in1}$ into equation (4-5) we find for the output currents:

$$I_{out+}=I_{in1}+\Delta I_{in1}\cdot\frac{\Delta I_{in2}}{I_{in2}} \tag{4-6}$$

$$I_{out-}=I_{in1}-\Delta I_{in1}\cdot\frac{\Delta I_{in2}}{I_{in2}}$$

With $\Delta I_{out}=I_{out+}-I_{out-}$ it simply follows for the differential output current that:

$$\Delta I_{out} = 2 \cdot \frac{\Delta I_{in1} \cdot \Delta I_{in2}}{I_{in2}} \tag{4-7}$$

So, using a small six-transistor circuit, we have obtained a differential output current which is proportional to the product of the two differential input currents scaled by the common-mode component of the second input current. From the circuit it can be found that the common-mode component of the output current is practically equal to the common-mode of the first input current.

The multiplication property is tested by means of PSPICE simulations. Fig.4-12 shows the simulation results. The first picture shows the

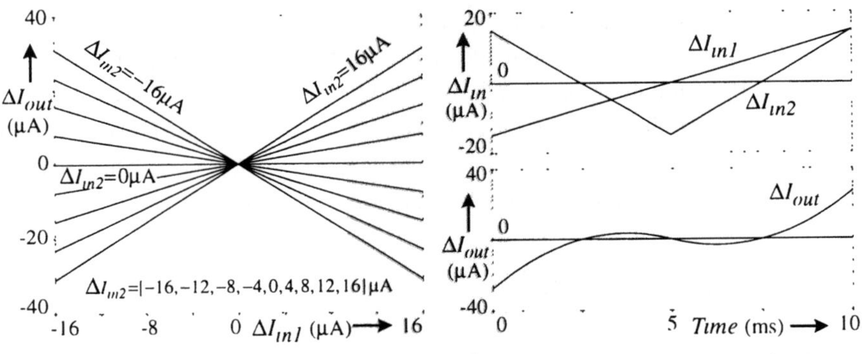

Fig. 4-12 *Simulation results showing the operation of the multiplier.*

differential output current as a function of one differential input current for different values of the other differential input current. The second picture shows the two differential input currents sweeping between plus and minus 16µA, and the resulting output current. Clearly, the multiplication works in all four quadrants, i.e. for negative and positive values of both differential input currents. The common-mode input currents were biased at 20µA.

4.5.3 Programmable current divider DAC

The programmable calibration coefficients are implemented in the form of so-called multiplying DACs. As explained in section 4.2.2, a DAC can be used to obtain a multiplication of an analog signal with a digital number. By connecting the DAC's digital input to a programmable digital storage register, the desired function is realized. Fig.4-13 shows the 8-bit DA-converter designed for differential current signals.

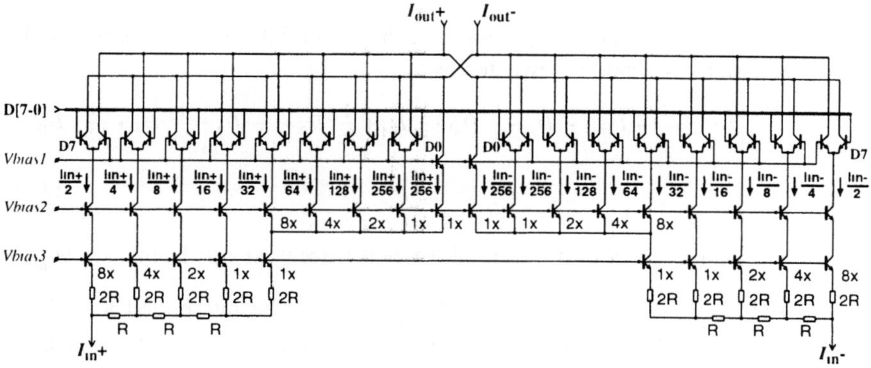

Fig. 4-13 *8-bit differential current-mode DAC.*

The currents on the positive input I_{in+} and on the negative input I_{in-} are divided into 8 binary-scaled parts, using cascode dividers [4]. Each binary part is switched to either the positive or the negative output by using a differential pair. The circuit is configured in such a way that the differential input current $\Delta I_{in}=(I_{in+}-I_{in-})$ is multiplied with the *two's complement* value of the 8-bit digital signal $D_{[7\text{-}0]}$ applied at the switches. This is obtained by reversing the connections of the differential pair for the most significant bit (MSB) D_7, and by connecting the remainder of the current division $\frac{1}{256}\Delta I_{in}$ in opposite phase to the output. When the base currents of the transistors are neglected, the following transfer for the differential current signal can easily be derived:

$$\frac{\Delta I_{out}}{\Delta I_{in}} = -\frac{1}{2}D_7 + \frac{1}{4}D_6 + \frac{1}{8}D_5 + \frac{1}{16}D_4 + \frac{1}{32}D_3 + \frac{1}{64}D_2 + \frac{1}{128}D_1 + \frac{1}{256}D_0 - \frac{1}{256} \tag{4-8}$$

in which the digital bits D_n can have a value of *+1* (*high*) or *-1* (*low*). The differential pair switches in the DAC in Fig.4-13 are all biased with one base to a reference voltage V_{bias1}. The other side of each switch either gets a lower or a higher voltage. As a differential voltage of 200mV is sufficient, the switches can be controlled by the output of a digital ECL (Emitter-Coupled Logic) circuit. The other bias voltages, V_{bias2} and V_{bias3}, need to be set at such a level that saturation of the transistors is prevented. V_{bias3} can be used to control the voltage level at the current inputs.

To allow a normal interpretation of the digital bits with values of *1* and *0*, equation (4-7) can be rewritten as:

$$\frac{\Delta I_{out}}{\Delta I_{in}} = -1 + \overline{D_7} + \frac{1}{2} D_6 + \frac{1}{4} D_5 + \frac{1}{8} D_4 + \frac{1}{16} D_3 + \frac{1}{32} D_2 + \frac{1}{64} D_1 + \frac{1}{128} D_0 \qquad (4\text{-}9)$$

The range of the differential output current spans from $-1\ \Delta I_{in}$ to $+\frac{127}{128}\ \Delta I_{in}$. The digital signals $D_{[7\text{-}0]}$ can now be interpreted as a digital *two's complement* encoded 8-bit number with value *d*. Which makes it possible to simply write the transfer as:

$$\Delta I_{out} = d \cdot \Delta I_{in} \qquad (4\text{-}10)$$

The digital number can have positive as well as negative polarity. This, naturally, also applies for the differential input current, thus, again, a four-quadrant multiplication is obtained. It is clear from the schematic that the common-mode output current is practically equal to the common-mode input current.

Because of transistor and resistor mismatch, the scaling in the binary dividers of the DAC is not perfect. This limits the resolution for which monotonicity can be guaranteed [4]. For this reason a low resolution of 8 bits was chosen for the experimental realization. Higher resolutions, 10 to 12 bit, would be possible when taking special measures for improved component matching. More advanced circuit techniques (e.g. dynamic element matching) are necessary when resolutions in the order of 12-16 bit are desired. This is not elaborated because it would complicate the design of the complete system considerably.

4.5.4 VI-converters and IV-converters

For the VI-conversion that is needed at the voltage inputs of the system, a standard cell VI-converter was available in the analog library. The schematic is shown in Fig.4-14. It originates from [4]. Transistors Q1 and

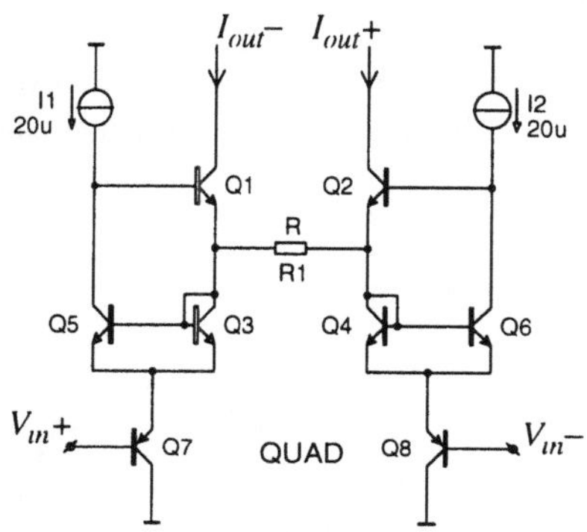

Fig. 4-14 *VI-converter used for converting the input voltages.*

Q2 deliver the output currents, while Q7+Q3 and Q8+Q4 perform a voltage following action from the input voltage to the resistor R1. The local feedback loops with Q1,Q3,Q5 and Q2,Q4,Q6 are constructed in such a way that the currents through Q3,Q7 and Q4,Q8 are, in first approximation, independent of the signal. The differential input voltage is copied exactly (besides offset) to the emitters of Q1 and Q2, across the feedback resistor R1. The differential output current, $\Delta I_{out}=I_{out+}-I_{out-}$, is twice as large as the current through the resistor, which is simply determined by the input voltage difference and the resistor value, neglecting the base currents of Q1 and Q2:

$$\Delta I_{out} = 2 \cdot \frac{V_{in+} - V_{in-}}{R} = 2 \cdot \frac{\Delta V_{in}}{R} \tag{4-11}$$

Because of the constant current through the voltage following transistors, the linearity of the circuit is considerably improved with respect to a normal degenerated differential pair. Another advantage of the circuit is that the input common-mode voltage can be biased on "ground-level"; one voltage input can be connected to the negative rail (ground), while the voltage input signal is applied at the other input.

For the current duplication blocks, see section 4.5.1, a combination of one IV-converter and two VI-converters is used. In fact, this combination forms a sort of current mirror, but then only for the differential current signal. The IV-converter turns the differential input current into a voltage which is then used by two VI-converters to generate two differential current signals.

Fig.4-15 shows both converters. The VI-converter, on the right, is just a differential pair with Darlington transistors and a degeneration resistor.

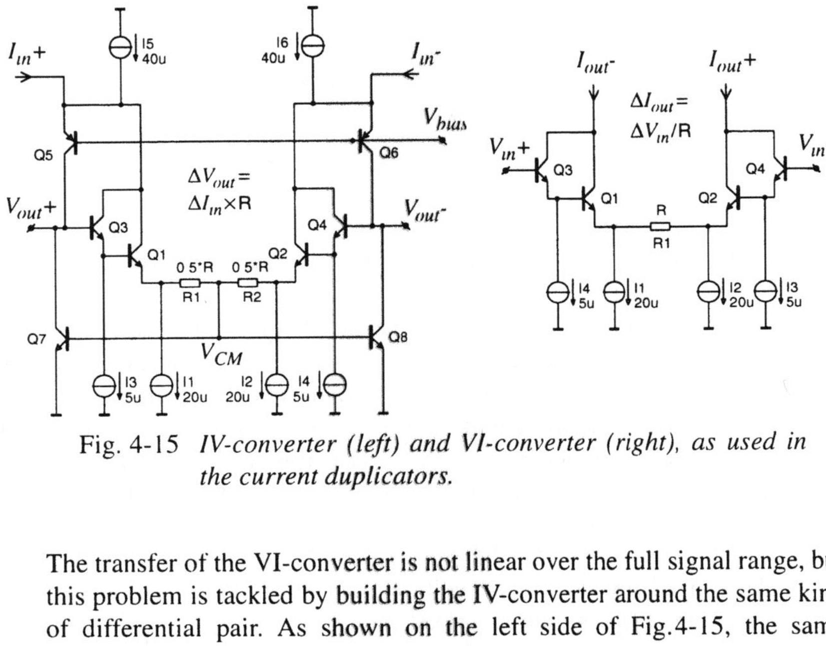

Fig. 4-15 *IV-converter (left) and VI-converter (right), as used in the current duplicators.*

The transfer of the VI-converter is not linear over the full signal range, but this problem is tackled by building the IV-converter around the same kind of differential pair. As shown on the left side of Fig.4-15, the same differential pair is used as a feedback from the output voltage to the input current. The loop is closed by the cascode transistors Q5 and Q6. Thus, the IV-converter has exactly the reciprocal transfer of the VI-converter. A common-mode control loop is made by splitting the resistor in two, and inserting Q7 and Q8. This is necessary to fix the common-mode output voltage level, independent of the common-mode input current.

The simulation result displayed in Fig.4-16 shows the IV-VI-transfer from differential current signal to a differential voltage, and then back to a differential current for the case that the IV-converter output is connected to the VI-converter input. The feedback resistors in the IV-converter

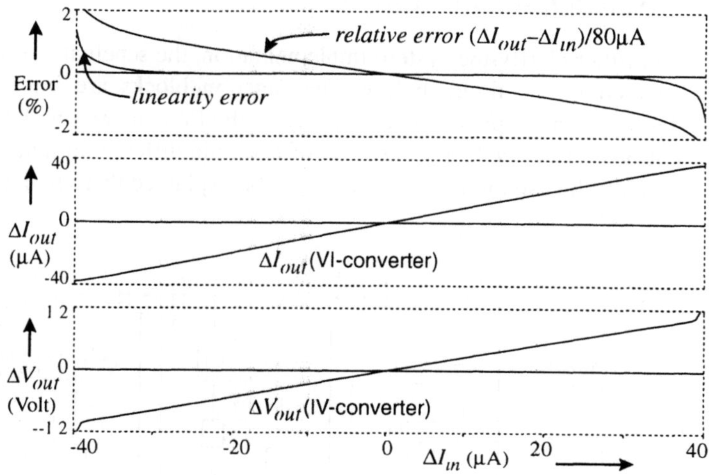

Fig. 4-16 *Total transfer of I-to-V and V-to-I conversion.*

(R1+R2) and in the VI-converter (R1) were given equal values (50 kΩ). The upper graph shows the relative error between differential input and output current, before and after a gain correction. We see that the linearity error in the current duplication will be small enough ($<<\pm0.2\%=$ 1/2 LSB for 8 bit) when the current range is limited to about [-30μA, 30μA].

In the current duplicator blocks used in Fig.4-10, two VI-converters are connected to the outputs of the IV-converter, so that two 'copies' of the differential current are obtained. By choosing different values for the feedback resistors, the output currents can be scaled up or down. This makes it possible to set the error-correction-range of each of the calibration steps.

The common-mode output current of the duplicator blocks can be adjusted independently of the input current. The common-mode (CM) voltage level at the current inputs is determined by the bias voltage at V_{bias} in Fig.4-15. In all described function blocks, the CM-voltage level at the outputs has to be higher than the CM-voltage level at the inputs, with the exception of the IV-converter shown in Fig.4-15. Here, the CM-level can go down. This is necessary to avoid a high 'stack' of function blocks and the need for a large supply voltage.

4.5.5 System implementation

To further clarify the system implementation, the schematic of Fig.4-10 is recalled in the figure below. The function blocks are implemented as described in the previous sections. Only the IV-converter at the output has not been described. It makes use of a simple differential amplifier which will also be used in *chapter 5* and will be explained there in detail.

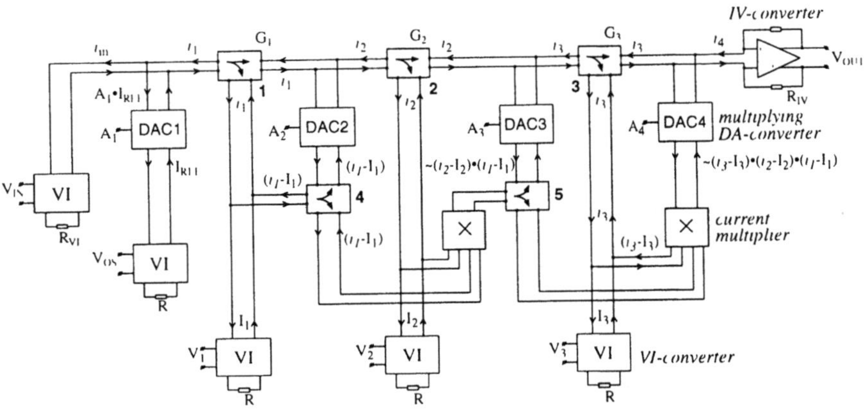

Fig. 4-17 *Programmable analog signal processor for progressive polynomial calibration.*

The system of Fig.4-17 is completed with the circuitry for the biasing of all the different analog blocks, and a digital circuit in Emitter-Coupled Logic (ECL), as presented in Fig.4-18.

The digital circuit consists of four 8-bit registers and a small serial interface which makes it possible to program each register separately. The digital outputs of the registers are connected directly, without level-shift, to the 8-bit digital inputs of the DACs. These digital signals are on ECL level which is equal to V_{CC} for *high* bits and $V_{CC}-V_{ECL}$ for *low* bits. V_{CC} is the positive supply voltage and V_{ECL} is the small differential voltage used in the ECL gates, approximately 300 to 400 mV. As a reference for the switches in the DACs, a voltage halfway *high* and *low*, $V_{CC}-{}^1/_2\, V_{ECL}$, is derived in the digital part and buffered to the analog part of the circuit.

The other bias voltages needed in the analog blocks are generated as shown in the figure. The bias voltages are chosen in such a way that

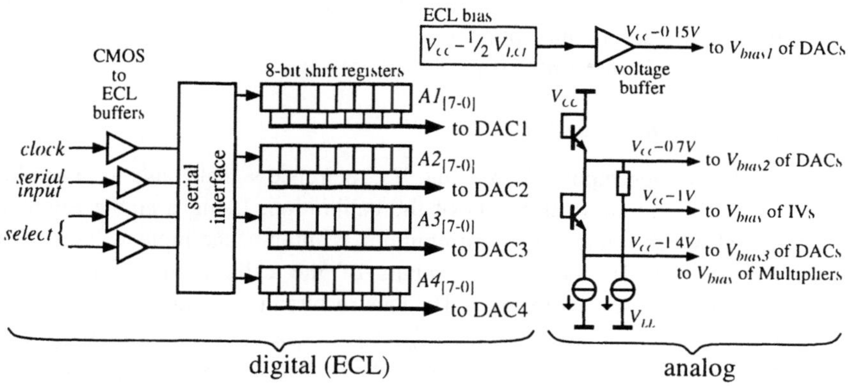

Fig. 4-18 *Additional circuit blocks for the analog signal processor.*

saturation of transistors is avoided when the blocks are connected as shown in Fig. 4-17, and the supply voltage is 4.5 Volt or higher.

The digital inputs of the serial interface are connected to small buffer circuits which convert digital input signals on CMOS-level [0-5 Volt] to signals on ECL-level [4.7-5 Volt]. This makes it easier to program and test the Integrated Circuit, using standard digital signals.

4.5.6 System operation

The circuit of Fig. 4-17 has a straightforward path through which the main signal flows: at the input the voltage signal is converted into a differential current, which then goes through the current duplicator blocks numbered 1 to 3, and at the output it is converted back into a differential voltage signal. The output currents of the DACs are added at several places in the forward path. When the calibration starts, all DACs are programmed to give a zero output current. As a result, the differential output voltage is directly proportional to the differential input voltage. The gain is determined by the ratio of the resistors in the input VI-converter and the output IV-converter, and the current scaling in the duplicators 1 to 3:

$$\frac{V_{OUT}}{V_{IN}} = G = 2 \cdot \frac{R_{\mathrm{IV}}}{R_{\mathrm{VI}}} \cdot G_1 \cdot G_2 \cdot G_3 \qquad (4\text{-}12)$$

The resistors R in Fig.4-17 are all equal, and the current duplicators 1 to 3 have scale factors one, thus the overall gain G is 2. However, by choosing different resistor ratios it is possible to obtain a larger gain.

Calibration Procedure

At the voltage inputs V_1, V_2, and V_3, reference signals have to be applied to fix the rotation set-points of the calibration. Using a stable reference signal which relates to the minimum/maximum of the input voltage range (sensor signal), the voltage inputs of the calibration circuit can be connected as shown in Fig.4-19. This results in calibration points for the input signal at respectively $V_{in}=V_1=-V_{reference}$, $V_{in}=V_2=+V_{reference}$, $V_{in}=V_3=0$ Volt. Through equation (4-12) the corresponding calibration points on the output voltage range can be found.

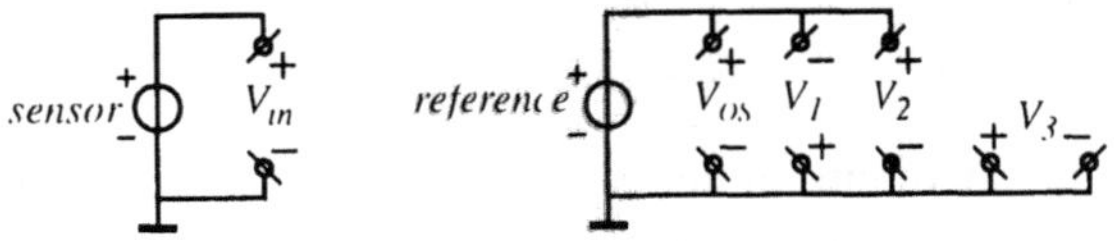

Fig. 4-19 *Connection of the voltage inputs of the calibration circuit to input and reference voltage sources.*

To follow the procedure, suppose that a sensor with a voltage output is connected to the input of the calibration circuit, and also a reference voltage is connected as shown in Fig.4-19. The combination of sensor and calibration circuit is then inserted in a test environment to calibrate the overall transfer.

The calibration starts by providing a first sensor reference signal (φ_{min}), which leads to an input voltage V_{IN} at the circuit. DAC1 is then programmed in such a way that the output voltage of the circuit obtains the desired value for the first measurement ($V_{OUT}=-G \cdot V_{reference}$). The required digital value can be calculated simply by taking two measurements of the output signal for two different values stored in the DAC (for example, 0 and $+^1/_2$ or $-^1/_2$), and interpolating the two measured output signals.

Then, a second input reference (φ_{max}) can be provided to the sensor and DAC2 can be programmed to obtain the desired output voltage for the second measurement ($V_{OUT}=+G \cdot V_{reference}$). Because of the system set-up, it will not be necessary to reprogram DAC1.

Then, a third calibration step is done to program DAC3, and a fourth step for DAC4, according to the calibration principle explained in section 3.3. After calibration, the combination of sensor and circuit has a signal transfer which resembles the desired transfer, and matches it exactly in the calibration points.

Simulation

By means of a PSPICE simulation, the circuit is tested for an input voltage which is swept through the input range. The input range is set to [-250 mV, +250 mV], V_{OS} to 250 mV, and the successive calibration points are set by voltages V_1= -250 mV, V_2= +250 mV, and V_3= 0 mV.

With R=25kΩ, the corresponding signal range for the *differential current* is approximately [-20 μA, +20 μA] and the full scale is 40 μA. The function blocks in the calibration circuit are biased accordingly for operation in class-A mode. Fig.4-20 shows the differential current signals at the inputs of the DACs as a function of the input voltage V_{IN}, when all DACs are set to zero.

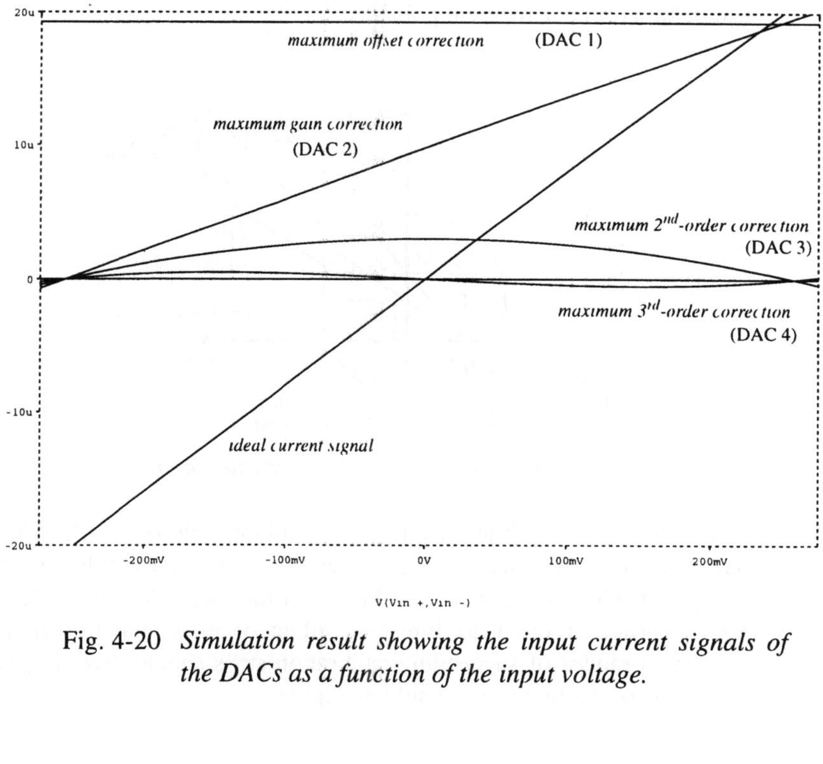

Fig. 4-20 *Simulation result showing the input current signals of the DACs as a function of the input voltage.*

The current signals in Fig.4-20 represent the maximum correction signals that can be applied for, respectively, offset, gain, and second-order and third-order linearity calibrations. In this case the maximum offset correction is almost ±50% of the full scale. This might be adjusted by changing the voltage applied at the V_{OS} input. The output currents of current duplicators 1 and 4 are scaled in such a way that the maximum gain correction signal is ±50% of the full-scale signal. The scaling in duplicators 2, 3, 4, and 5 is such that the second-order correction is maximum ±8%, and the third-order approximately ±2% of the full-scale output. All the corrections can be done with an 8-bit resolution.

Calibration Errors

The limited resolution in the calibration of the offset leaves a remaining offset error in the calibration point: in the worst case, it could be plus or minus a half *LSB* (Least Significant Bit) of the offset calibration range. The gain correction at the next calibration step makes use of the error signal, which is now not exactly zero in the previous calibration point. This can have a negative effect on the remaining offset error, as illustrated in Fig.4-21. The transfer will not rotate around the intended point in the origin but around a different point.

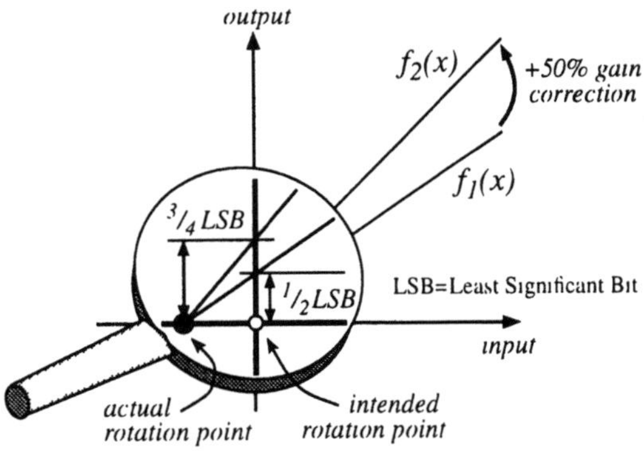

Fig. 4-21 *Effect of an error in the first rotation point.*

When the offset calibration error was calibrated down to $^1/_2$ *LSB*, and, next, the gain is increased by 50%, then the offset error will be magnified to $^3/_4$ *LSB*. The same effect also applies for the succeeding steps, but with less severity. In order to limit the effect of error magnification, the correction ranges of succeeding calibration steps should decrease, or the resolution of the first steps should be higher.

Fig. 4-22 shows the error curve after calibration for a worst-case example. Four calibration steps were done at calibration points corresponding to -250 mV, +250 mV, 0 mV, and -125 mV on the input range. The considered

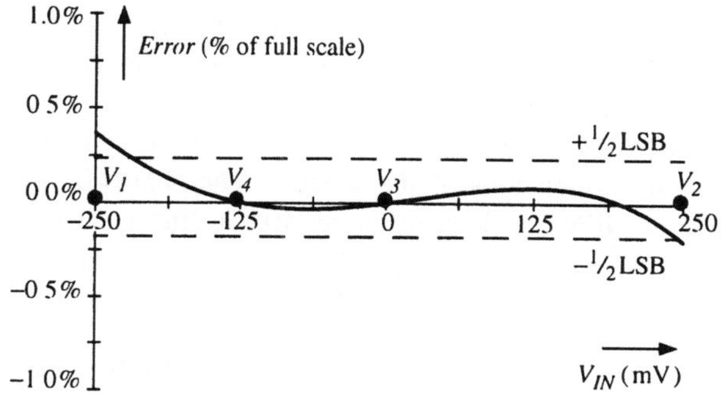

Fig. 4-22 *Error after four calibration steps with 8-bit resolution.*

example curve required almost maximum gain and linearity corrections. Initially, at the first calibration step, the offset error at V_{IN}=-250 mV was reduced to $^1/_2$ *LSB*, or 0.2% ($=^1/_2 \times 2^{-8} \times 100\%$) of the full-scale signal. But after the gain and linearity correction it increased to approximately 0.35%. The gain calibration itself left an error at V_{IN}=250 mV of $^1/_2$ *LSB*, or -0.2%. Because of the much smaller signal range for the linearity corrections, the same 8-bit resolution leads to a much smaller error in the third and fourth point. The resulting integral error over the input range is relatively small, despite the larger error values at the edges of the range.

To keep the error within the $\pm^1/_2$ *LSB* band, the resolution of the DACs should be adapted to the correction ranges for each calibration step. For practical reasons the use of the same 8-bit DAC and 8-bit registers for all four steps was preferred in the presented realization.

4.5.7 Realization

For the digital part of the circuit, the developed ECL standard cell library for the DIMES-01 process was used. The layout of the digital circuit has been generated automatically by the computer layout program L-Edit.

Correspondingly, the presented analog function blocks have been laid out as analog standard cells, so that they could also be used in the automatic 'Place and Route' routine of the same layout program. The standard cells on transistor level were laid out manually. Where necessary, transistor pairs were placed in a cross-coupled double pair configuration (quad) to minimize the offset. Where the ratio of feedback resistors was important, the resistors have been put close together in a standard cell to improve matching. On the system level the connections were made automatically. The analog and the digital part also had to be connected by hand. Fig.4-23 shows a photograph of the resulting chip, which measures approximately $5 \times 5.5\,\mathrm{mm}^2$. This seems large, but it has to be considered relative to the minimum dimensions of the process. The DIMES-01 process is not very dense; an NPN transistor measures $40 \times 56\,\mu\mathrm{m}^2$, and a metal 2-metal 1 via is $12 \times 12\,\mu\mathrm{m}^2$. In a modern high-density bipolar process, the square chip size would scale down by a factor of four at least.

Fig. 4-23 *Chip photograph of the realized calibration circuit.*

The functionality of the integrated circuit has been tested in a measurement set-up as illustrated in Fig.4-24. Only the supply of 5 Volt is not shown. A constant voltage signal from a DC source is applied at the reference inputs in such a way that V_l=−250mV and V_{OS}=V_2=250mV. The third input is short-circuited so that V_3=0mV. Using a signal generator, a low-frequency triangle-wave signal with an amplitude of approximately 300mV is applied at the input V_{IN}. The CM-level of the input voltage was set at the ground rail (0V). From a PC using the data-acquisition program LabView, the digital registers which control the DACs could be programmed via the four digital inputs of the circuit. The resulting differential output signal was observed on an HP digital storage oscilloscope. Because of a gain of two, the output signal is twice as large as the input signal. However, the output voltage is limited by the circuit's signal range at approximately +600mV and –600mV.

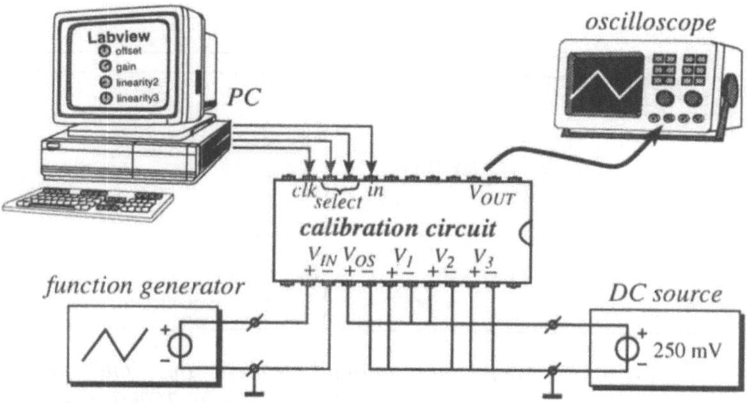

Fig. 4-24 *Measurement set-up.*

The triangular signal observed on the oscilloscope shows the transfer of the circuit for the considered input range. The four registers of the calibration circuit were programmed with different values, and the output traces were stored on the scope. This made it possible to display several traces simultaneously on the oscilloscope, which is shown in the scope plots of Fig.4-25. The first picture shows the traces for different values programmed in the first register for the offset. The second picture shows the traces for different values stored in the second register for the gain. Idem ditto for the third and the fourth picture.

The pictures clearly show the rotation points in the output range at -500mV, +500mV, and 0V. The first three calibration points thus have to be chosen at those output values. The fourth and last calibration point can be chosen freely, but the best choice for a third-order linearity correction is around plus or minus 250mV.

Because the analog signal goes through a cascade of local-feedback amplifiers (IV-converters and VI-converters) with a relatively large bandwidth, the bandwidth of the complete circuit can be large as well. So,

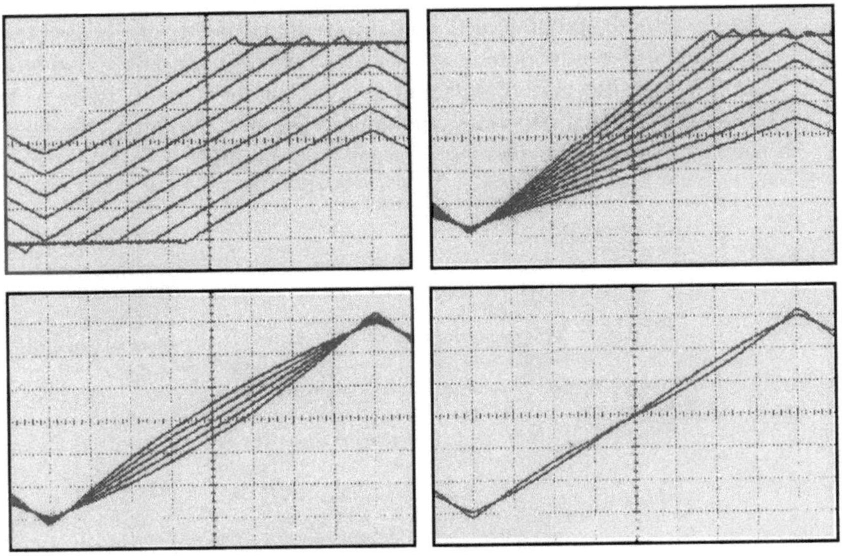

Fig. 4-25 *Measured output traces on the oscilloscope (x-axes: 5ms per division, y-axes: 200mV per division).*

after the sensor transfer is calibrated, using low-frequency or static reference signals, the correction also works for analog signals of high frequencies. Depending on the power invested in the bias currents of the different circuit blocks, bandwidths in the order of MHz can be achieved. To reduce output noise, the bandwidth of the realized circuit has been limited on purpose with a low-pass filter to approximately 10kHz. The specifications are summarized in the table below.

4.6 Conclusion

Several programmable analog calibration circuits, based on the use of Digital-to-Analog Converters, have been presented. For uncomplicated sensor calibrations requiring a limited resolution for the correction factors (8-10 bit), these circuits can be advantageous. With respect to conventional sensor trimming techniques, the advantage lies in the possibility to digitally program the calibration and hence facilitate automation. Furthermore, digitizing of the sensor signal can be avoided,

Table 4-1 *Specifications of the analog calibration circuit.*

Description	Value	Unit
Supply Voltage Vcc[a]	4.5-12	V
Current Consumption A=Analog, D=Digital	$1.2^A + 1.6^D$	mA
Input CM-range (forVcc=5V)	−0.3-0.6	V
Maximum differential input signal	±450	mV
Nominal Gain	2	V/V
Max. Offset Correction[b]	±50% of FS, 8 bit	
Max. Gain Correction	±50% of FS, 8 bit	
Max. 2nd-order Linearity Correction	±10% of FS, 8 bit	
Max. 3rd-order Linearity Correction	±2% of FS, 8 bit	
Max. Error after calibration (peak value)	±0.4% of FS[c]	
Conditions: Vcc=5 Volt, V_{OS}=$-V_1$=V_2=250mV, V_3=0V, FS_{in}=0.5V, FS_{out}=1V		

a. Provided that the digital inputs can be scaled up accordingly

b. Adjustable with V_{OS}

c. Provided that the nonlinearity of the original signal fits within the correction range.

and thus the analog calibration option offers the possibility to maintain the bandwidth and dynamic range obtainable with compact analog circuitry.

The polynomial linearization method presented in the previous chapter can be worked out systematically in a programmable analog calibration and linearization circuit. The implementation of such a circuit was elaborated and demonstrated by a bipolar chip realization. However, the implementation in such analog circuitry is feasible only for a small number of calibration steps or for sensors requiring a high bandwidth.

References

[1] D. Tandeske, *Pressure Sensors: selection and application*, Marcel Dekker Inc., New York, 1991.

[2] J. Bryzek, K. Petersen, J.R. Mallon, L. Christel, and F. Pourahmadi, *Silicon Sensors and Microstructures*, NovaSensor, Fremont, USA, 1990.

[3] P. Ciureanu and S. Middelhoek, *Thin Film Resistive Sensors*, Institute of Physics Publishing, Bristol, 1992.

[4] R. v.d. Plassche, *Integrated Analog-to-Digital and Digital-to-Analog Converters*, Kluwer Academic Publishers, Boston/Dordrecht/ London, 1994.

[5] A.B. Carlson, *Communication Systems: an Introduction to Signals and Noise in Electrical Communication*, McGraw-Hill Book Co., Singapore, 1986.

[6] J.H. Huijsing, F.R. Riedijk, and G. van der Horn, "Developments in integrated smart sensors", *Sensors and Actuators A*, Vol.43, 1994, pp.276-288.

[7] G. v.d. Horn and J.H. Huijsing, "Integrated smart sensor calibration", *J. Integrated Analog Circuits and Signal Processing*, Vol.14, No.3, November 1997 (*to be published*).

[8] G. v.d. Horn and J.H. Huijsing, "Programmable analog signal processor for polynomial calibration", *Proceedings ESSCIRC'96*, Neuchatel, Switzerland, 1996, pp.400-403.

[9] B. Gilbert, "A precision four-quadrant multiplier with sub-nanosecond response", *IEEE Journal of Solid-State Circuits*, Vol. SC-3, 1968, pp.365-373.

[10] C. Toumazou, F.J. Lidgey, and D.G. Haigh, edits., *Analogue IC design: the current-mode approach*, Peter Peregrinus Ltd., London, 1990.

Calibration using Sigma-Delta Analog-to-Digital Conversion 5

5.1 Introduction

In section 4.4 the use of pulse modulation techniques to control the signal transfer of a smart sensor was already suggested. The pulsestream or bitstream modulation can be combined nicely with sigma-delta bitstream techniques. A digital sigma-delta (noise shaper) may be used to generate programmable pulsestreams or bitstreams which modulate the analog signal, and a sigma-delta AD-converter is used to convert the modulated analog sensor signal into a digital bitstream.

Before presenting the application of this technique for smart sensor calibration, the explanation of the sigma-delta converters of section 2.4 will be extended to the frequency domain in section 5.2. In section 5.3 we will then explain the principle of the bitstream modulation, and the effects on the transfer and the performance of the AD-conversion. Implementation of a first-order sigma-delta converter in bipolar circuits will be presented in section 5.4. In section 5.5, we will present a smart temperature sensor with programmable calibration as an example for the application of the technique. Realizations of such smart temperature sensors, including one with a bus-interface, will be shown in section 5.5.

5.2 Sigma-delta AD-converters

For clarity, we recall the differential configuration of a first-order continuous-time sigma-delta AD-converter in Fig.5-1. Instead of using two switched reference currents, as in Fig.2-14, they have been assembled in one current source of twice the reference value. This is possible provided that the differential amplifier has an input common-mode (CM) control which takes care of the CM-component of the input current. The converter further consists of an integrator, which can be regarded as a first-order low-pass filter, and a clocked comparator, which can be viewed as a combination of a 1-bit quantizer and a clocked gate.

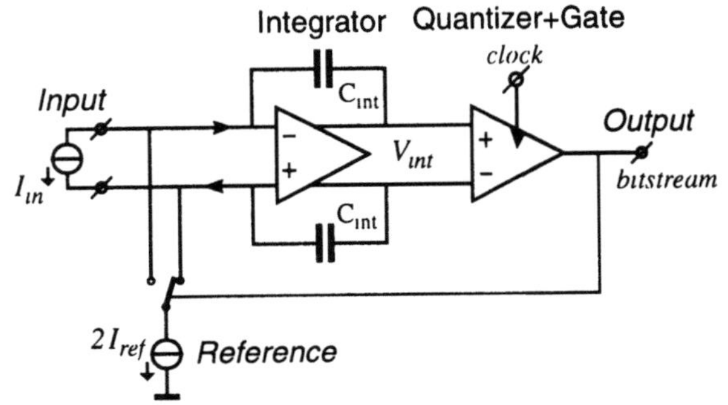

Fig. 5-1 *Differential sigma-delta converter configuration.*

Despite the quantization accuracy of only 1-bit, a high resolution conversion can yet be obtained on the basis of two well-known principles: oversampling and noise shaping. The oversampling is achieved by running the clock at a frequency much higher than the Nyquist sample frequency. The noise shaping is obtained by the low-pass filter function in the loop.

A high resolution multi-bit digital output can be obtained by digitally processing a long series of output bits. This is referred to as decimation. These digital output samples become available at a sample rate much lower than the original sample frequency, hence the often-used terms downsampling and decimation filter. Similar to the oversampling ratio, see equation (2-5), we can define a downsampling ratio *DSR* as the ratio of the original sample frequency f_s, and the decimated sample frequency f_{ds}:

$$DSR = \frac{f_s}{f_{ds}} \tag{5-1}$$

Usually, the maximum down-sample ratio *DSR=OSR* is chosen, and the decimated digital output samples are available at the Nyquist rate again.

Fig. 5-2 explains the complete signal processing schematically. As the 1-bit serial output of the sigma-delta converter is a very convenient sensor output signal, the digital filter is usually *not* incorporated in the smart sensor design. Instead, it may be implemented in software on the system's computer or microcontroller. Or, it can be implemented in digital hardware on a separate chip which is, for example, connected to the sensor bus.

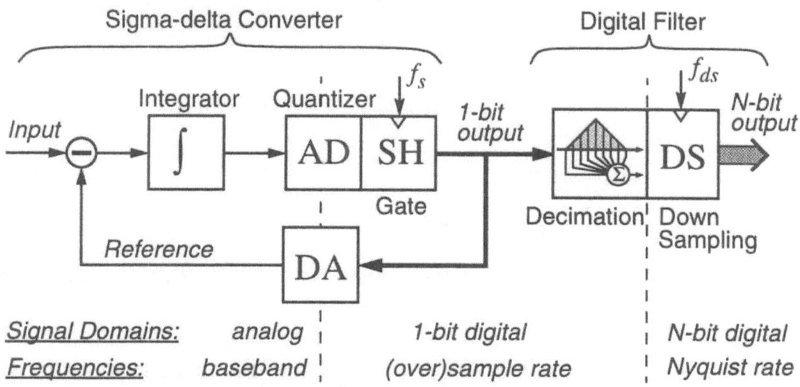

Fig. 5-2 *Signal processing chain of a sigma-delta AD-converter.*

The three principles involved in the sigma-delta type of AD-conversion will be explained shortly and intuitively. More elaborate discussions can be found in [1,2,3,4].

5.2.1 Principle of oversampling

The effect of quantization is often studied under the practical assumption that the probability of the input signal is distributed uniformly between the quantization levels. The quantization error is then distributed uniformly between $-{}^1/_2\,LSB$ and $+{}^1/_2\,LSB$, with the Least Significant Bit representing the smallest quantization step $q_s = 1\,LSB$. By integration of the square error over that range, the average power of the quantization error can be estimated as [5]:

$$\varepsilon^2 = \frac{1}{12} q_s^2 = \frac{LSB^2}{12} \tag{5-2}$$

The error introduced by the quantization is then considered as an additional noise source. It is further assumed to be uncorrelated to the sampled input signal. The noise will thus be spread evenly over a frequency band equal to half the sample frequency, as shown in Fig.5-3.

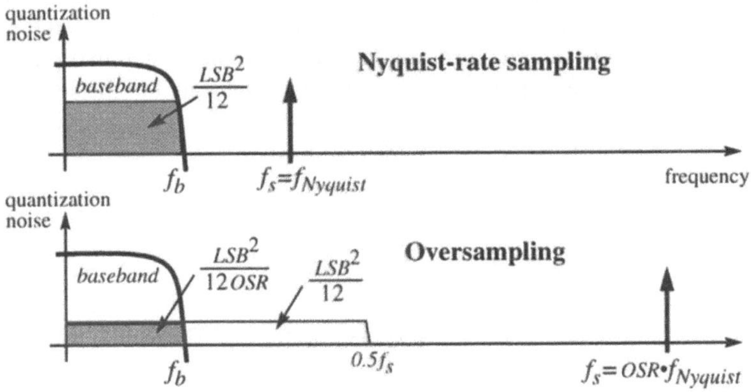

Fig. 5-3 *Effect of oversampling on the quantization noise in the baseband.*

When sampling at the Nyquist rate, all the quantization noise falls in the baseband. However, when oversampling with a certain ratio OSR, the quantization error is spread over a much larger bandwidth [2,6,7,8]. The part outside the baseband can be removed by a filter and does not affect the attainable signal-to-noise ratio. The baseband noise power is thus reduced by a factor equal to the OSR. Accordingly, the signal-to-noise ratio is improved with 3dB for each doubling of the sample frequency. The improvement is given by:

$$F_{0,\,dB} = -20\log\sqrt{OSR} \tag{5-3}$$

5.2.2 Principle of noise shaping

To explain the principle of noise shaping a discrete-time model of the sigma-delta converter is most often used [1,2]. Even when using a continuous integrator rather than the inherently time-discrete switched-capacitor integrator. The model remains valid because the comparator processes the integrator output at discrete time steps only. For a linear integrator, the output voltage at discrete time steps, $t = n\mathrm{T_{clk}}$, is determined by the previous output and the input current:

$$V_{out}[n] = V_{out}[n-1] + I_{int} \bullet \mathrm{T_{clk}} / \mathrm{C_{int}}. \tag{5-4}$$

A general model of a first-order converter, also called noise-shape coder, is shown in Fig. 5-4. The input signal $X(z)$ is normalized with respect to the reference of the converter, so that the reference signal could be omitted from the model.

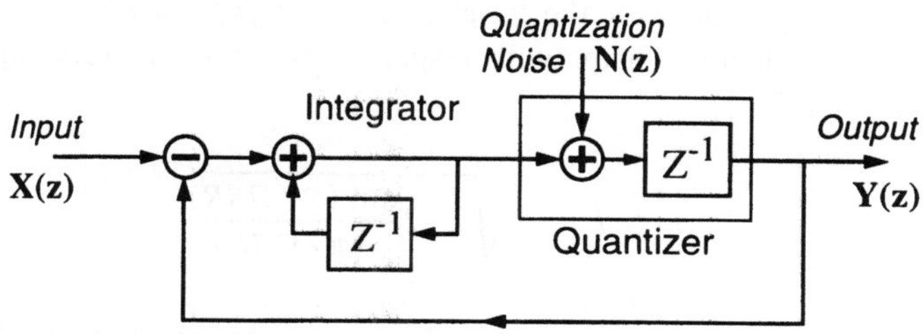

Fig. 5-4 *Discrete-time model of a first-order sigma-delta converter.*

The quantizer is modeled by the noise source $N(z)$ and the gate by a unit delay z^{-1}. As we will see, the transfer of the integrator is essential for the noise shape. It is given by:

$$H(z) = \frac{1}{1 - z^{-1}} \tag{5-5}$$

Using the figure, the transfer from the signal input $X(z)$ and the noise input $N(z)$ to the quantized output $Y(z)$ can easily be found, and expressed as:

$$Y(z) = z^{-1} \cdot X(z) + (1 - z^{-1}) \cdot N(z) \tag{5-6}$$

Whereas the input signal is only delayed (phase-shift), the quantization noise is shaped by the factor $(1\text{-}z^{-1})$. The effect on the noise magnitude characteristic is obtained by calculating $|1\text{-}z^{-1}|^2$ with substitution of

$z=e^{-j\omega\tau}$, $\omega=2\pi f$, and $\tau=T_{clk}=1/f_s$. As a result, we find the following function for the shape of the quantization noise power:

$$|N(z)|^2 = 2\left(1 - \cos\frac{2\pi f}{f_s}\right) \tag{5-7}$$

The function shows that the quantization noise is suppressed at low frequencies, in the baseband, and elevated at higher frequencies, around half the sample frequency. The noise-shape function is shown in Fig.5-6.

By integration over the baseband, $f=0$ to $f=f_b=\frac{1}{2}f_s/\mathrm{OSR}$, we can find a noise reduction factor with respect to the normal oversampled uniform noise [2]:

$$F_1 = \sqrt{2-2\frac{\sin(\pi/OSR)}{(\pi/OSR)}} \tag{5-8}$$

For sufficiently large oversampling ratios, this can be worked out, using a Taylor expansion for $\sin(\pi/OSR)$. Expressed as an *improvement* factor for the signal-to-noise ratio in dB, it is given by:

$$F_{1,dB} = -20\log\sqrt{3}\left(\frac{OSR}{\pi}\right) \tag{5-9}$$

For each doubling of the sample frequency, a 6dB improvement in the signal-to-noise ratio in the baseband may be obtained by the noise shaping, in addition to the 3dB improvement obtained by the oversampling action (equation (5-3)).

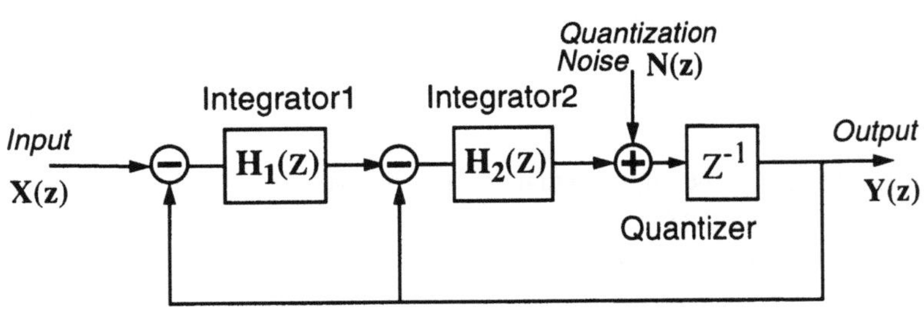

Fig. 5-5 *Possible model for a second-order sigma-delta configuration.*

Further improvement can be obtained with higher order noise-shape functions. With higher order noise shapers, the number of possible configurations also increases, and the stability becomes an issue for concern [9,..., 14]. We will restrict ourselves to a second-order sigma-delta modulator as modeled in Fig.5-5.

With integrator functions $H_1(z)$ and $H_2(z)$ as given by equation (5-5), a transfer of the quantization noise $N(z)$ to the output $Y(z)$ proportional to $(1-z^{-1})^2$ is obtained. The corresponding amplitude characteristic gives the following noise-shape function:

$$|N(z)|^2 = |(1-z^{-1})^2|^2 = 6 - 8\cos\frac{2\pi f}{f_s} + 2\cos\frac{4\pi f}{f_s} \qquad (5\text{-}10)$$

Both the first-order and second-order noise-shape function are shown in Fig.5-6. Clearly, the second-order function suppresses the quantization noise in the baseband even further.

By integration of the equation (5-10) over the baseband, we find the noise reduction factor, again with respect to the uniform case, expressed as a function of the oversampling ratio [2]:

$$F_2 = \sqrt{6 - 8\frac{\sin(\pi/OSR)}{(\pi/OSR)} + \frac{\sin(\pi/OSR)}{(\pi/OSR)}} \qquad (5\text{-}11)$$

By means of Taylor expansions this can also be simplified. The improvement in the signal-to-noise ratio can then be expressed in dB by:

$$F_{2,dB} = -20\log\sqrt{5}\left(\frac{OSR}{\pi}\right)^2 \qquad (5\text{-}12)$$

Thus a 12dB improvement of the signal-to-noise ratio for each doubling of the sampling frequency can be achieved with the second-order noise shaping, in addition to the 3dB for the oversampling action.

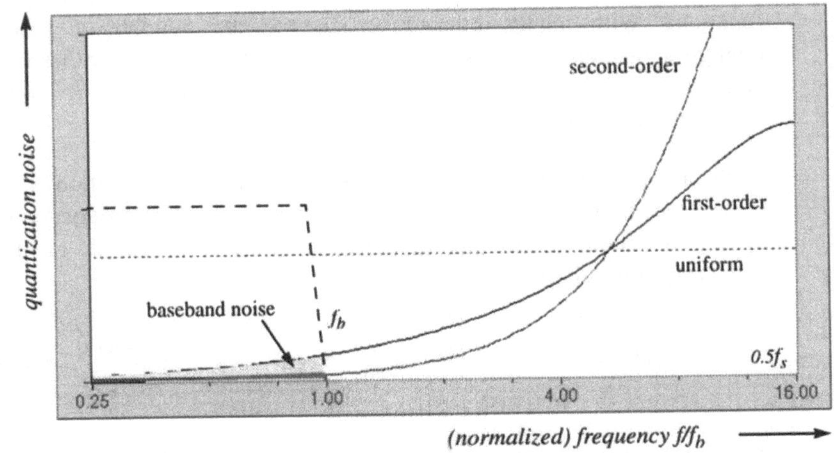

Fig. 5-6 *Noise-shape functions of a first-order and second-order sigma-delta modulator for OSR=16.*

The above described method can be applied for a generalized n-bit noise shaper (sigma-delta) of order L, to find a signal-to-(quantization)noise ratio that can be estimated by [3]:

$$\frac{S}{N} = \frac{q_s 2^n / 2\sqrt{2}}{q_s / \sqrt{12}} \cdot F_0 \cdot F_L = 2^n \sqrt{3\pi\left(L + \frac{1}{2}\right)} \cdot \left(\frac{OSR}{\pi}\right)^{L + \frac{1}{2}} \qquad (5\text{-}13)$$

A maximum sine-wave signal with a peak-to-peak value of $2^n \cdot q_s$ was assumed. For low resolution (low n) noise-shaping coders, however, van de Plassche proposes a correction factor for the maximum signal amplitude [2]. For a 'one-bitter', the maximum peak-to-peak value of the sine wave signal would be $4q_s/\pi$ instead of $2q_s$.

In practice, the quantization noise of sigma-delta converters does not follow the explained noise-shape function smoothly. The first-order modulator in particular shows very discrete spectrum components, which relate directly to the input and the clock signal. These spectrum components correspond to so-called limit cycles which occur when a bitstream pattern of limited length repeats itself continuously. For DC input signals the behavior is completely deterministic.

Fig. 5-7 shows the spectrum of the bitstream output of a first-order sigma-delta converter with a -3dB sine-wave input signal at $f=f_b$, and one hundred times oversampling. It was obtained by means of a simulation in the data-acquisition program, LabView. The figure also indicates the theoretical noise-shape function.

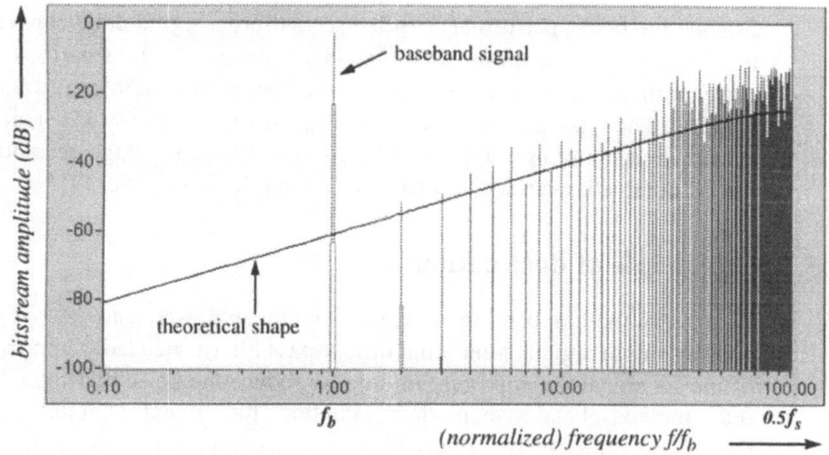

Fig. 5-7 *Frequency spectrum of a first-order sigma-delta bitstream output, for* OSR=100.

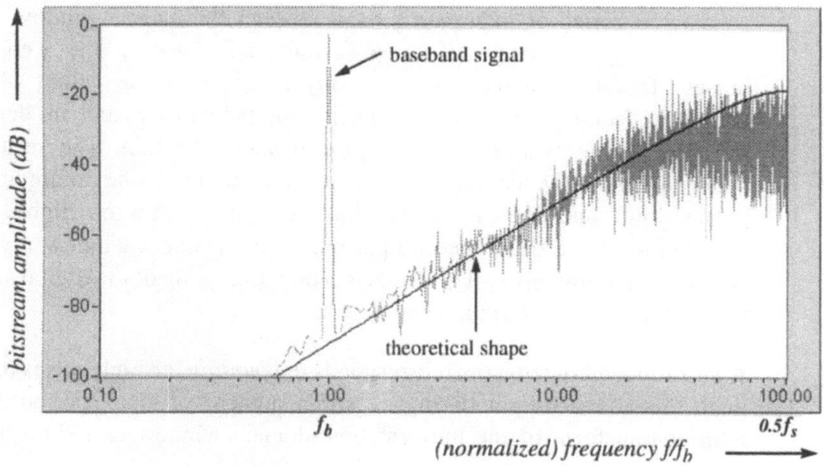

Fig. 5-8 *Frequency spectrum of a second-order sigma-delta bitstream output, for* OSR=100.

Fig. 5-8 shows the output spectrum for a second-order modulator, under certain conditions, and the theoretical noise-shape function. The spectrum clearly resembles the theoretical noise-shape function much better than in case of the first-order modulator. The quantization error is decorrelated more from the signal, and, hence, the assumption of *white* quantization noise is more applicable.

Despite the better performance of the second-order sigma-delta converter, the first-order converter is sometimes preferred in smart sensor configurations because of the simple implementations and a guaranteed stability. The typically low bandwidth signals of sensors (<1 kHz) allows a large oversampling ratio (>1000) so that sufficiently high resolutions can be obtained, even with a first-order sigma-delta converter [4].

5.2.3 Principle of decimation

To retrieve the baseband input signal from the high-frequency (HF) digital one-bit output signal, some kind of low-pass filter is needed. Suppression of the HF signals is important in order to reduce the noise that folds back into the baseband when downsampling the signal. Digital signal processing offers many advanced filtering techniques [15, 16, 17], but we restrict ourselves to the simplest technique used for bitstream decimation.

Each decimated output sample is obtained by processing the bitstream output for a certain number of clock periods, equal to the *DSR*. The considered series of bitstream pulses, as seen through a 'window', are processed by applying weighting factors and summing the weighted pulses. This is explained schematically in Fig. 5-9. The values of the weighting factors depend on their position in the window, and are defined by a weighting function, or so-called window function. The resulting decimated output value must be interpreted in relation to the minimum (all pulses low) and the maximum filter output (all pulses high). The calculation of the decimated output occurs each time a window is filled with fresh output pulses. This leads to a downsampling of the digital signal by a factor *DSR* equal to the window size.

A *uniform window* function corresponds to *counting* the high pulses (ones) in the bitstream output during a certain number of clock periods. The Fourier transform of the uniform time-domain window is a sinc-shaped low-pass function, as shown in Fig. 5-10. Hence the name *sinc filter* is used for this simple first-order decimation filter.

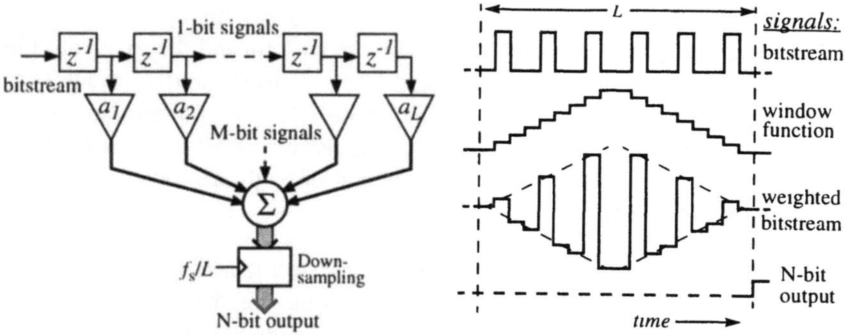

Fig. 5-9 *Application of window functions for decimation of the sigma-delta bitstream.*

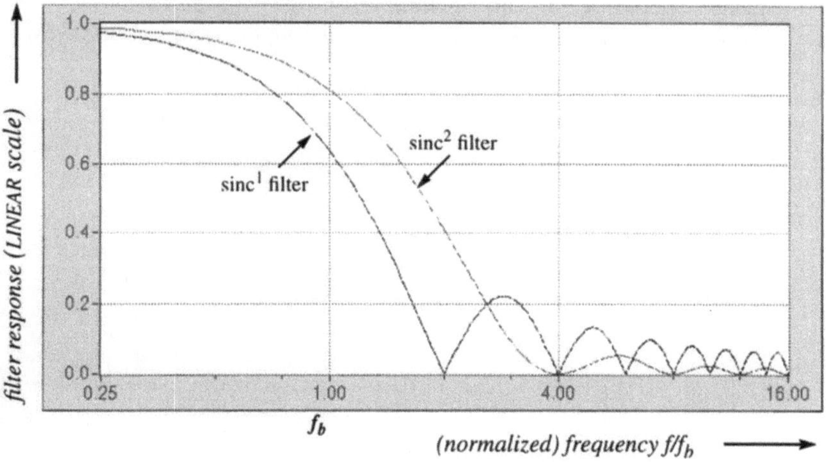

Fig. 5-10 *Frequency transfer functions of decimation filters with uniform and triangular window (of the same size).*

A *triangular window* function, as explained in Fig. 5-9, corresponds to a low-pass filter function with a sinc-square characteristic. The transfer of such a second-order or *sinc² filter* is also shown in Fig. 5-10. For both cases the same window size was used. Clearly the suppression of the higher frequencies by the sinc²-filter is stronger than for the sinc-filter.[1]

Steeper filter shapes can be obtained with more advanced window functions, such as Hanning, Hamming, Blackman, etc. [18]. Theoretically, a sinc-shaped time-window (with infinite window length) would realize the ideal uniform baseband filter in the frequency domain. In contrast to the advanced windows or filters, both the uniform and the triangular filter are very easy to implement in hardware, using digital counters and adders [15], as will be shown in section 5.4. The configuration is then somewhat different as the FIR (finite impulse response) structure explained in Fig. 5-9, in order to reduce the amount of hardware (imagine the delay line for large *OSR*).

If the order of the decimation filter is one higher than the order of the noise shaper, it provides enough digital resolution for the obtainable signal-to-noise ratio. The spectrum of the digital output signal of a first-order sigma-delta converter with a sine-wave input signal, after decimation with a triangular window function is shown in Fig. 5-11. The filter shape, also shown in the figure, clearly suppresses the HF quantization noise. The peaks appearing in the spectrum are the higher harmonics of the square-wave digital output signal representing the sampled sine-wave.

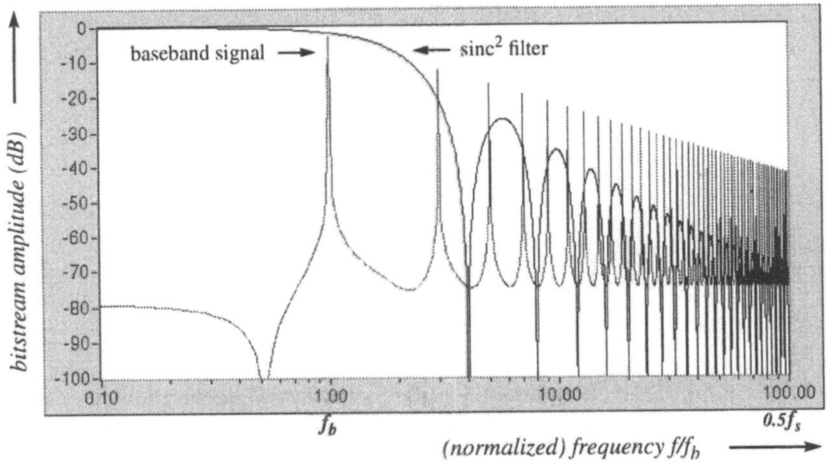

Fig. 5-11 *Output spectrum of decimated first-order sigma-delta output with OSR=100 and a sinc^2 filter with DSR=100.*

[1] One may recognize that when a rectangular window function is used in a moving-average filter (the average is computed every clock period), this corresponds to a triangular-shaped weighting function for a window that is twice as large.

5.2.4 Transfer from analog input to digital output

Now that we have explained all the processing required for the sigma-delta analog-to-digital conversion, we can show the input-output transfer. Fig.5-12 shows the transfer of a differential first-order sigma-delta converter in combination with a decimation filter based on a triangular window, using an oversampling ratio and window size of 128.

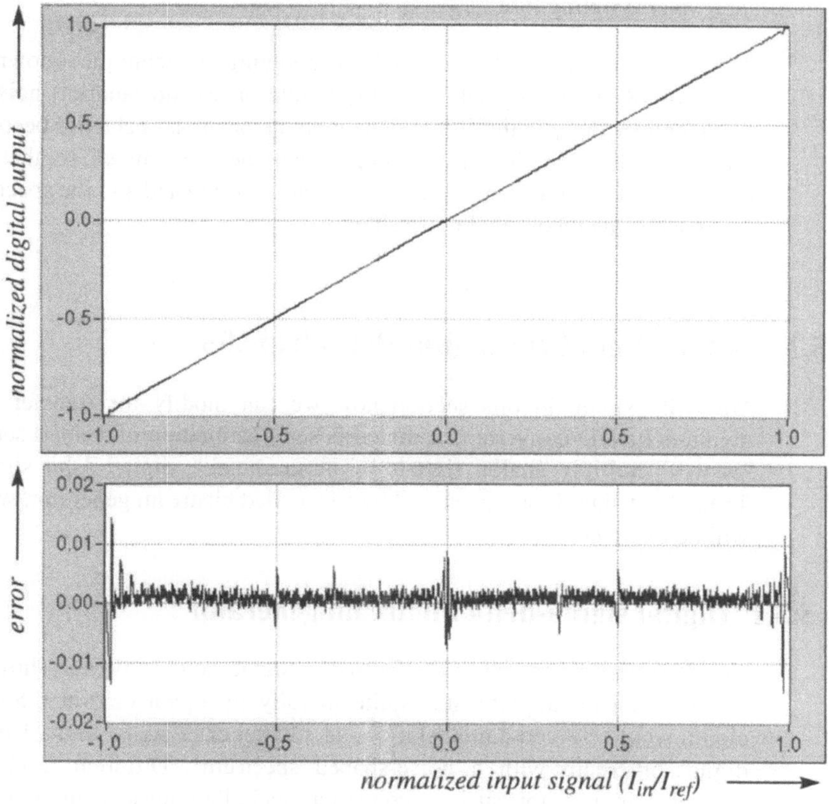

Fig. 5-12 *Transfer and error of a first-order sigma-delta converter using a sinc² filter for normalized signals and OSR=128.*

The input signal is normalized with respect to the sigma-delta reference, and the output signal is normalized with respect to maximum and minimum digital output.

In most parts of the input range the error is quite small, in accordance with expectations based on equation (5-13) (S/N≈2000). However, for input signals just around basic fractions of the reference, such as -0.5, -0.33, 0, 0.33, and 0.5, the error can get much larger. Around zero the error is approximately as large as $1/OSR$, which is due to long limit cycles occurring for small non-zero signals. These result in very low frequency components in the bitstream output, which can not be filtered out by the decimation filter. The error must be reduced sufficiently by choosing a large oversampling ratio.

The peak errors can be reduced by applying a technique known as *dithering* [8,19,20]. By adding some (white or pseudo-random) noise at the quantizer input in the sigma-delta loop the bitstream behavior becomes more chaotic. The limit cycle lengths will be randomized so that the discrete peaks in the frequency spectrum are smoothened and the spectrum approaches the theoretical noise shape.

5.3 Alteration of the sigma-delta transfer

We will explain in this section how we can modify the transfer of a sigma-delta AD-converter for the purpose of calibration of a smart sensor. We will multiply analog signals by programmable digital 1-bit signals. These 1-bit signals are generated by a so-called bitstream generator, which will be explained first.

5.3.1 Digital sigma-delta bitstream generator

The discrete-time model of the first-order sigma-delta converter, shown in Fig.5-4, can be implemented quite literally in digital circuitry. Such a circuit would convert a multi-bit digital number or LF signal into a 1-bit HF digital bitstream with a noise-shaped spectrum[1]. Different terms are presently in use for such a converter, including noise-shaping coder, sigma-delta (de)modulator, sigma-delta DA-converter, and fractional-N generator. As for the sigma-delta AD-converter, there are also many

[1] The same technique is applied for inexpensive high-accuracy DA-converters, for example the well-known 'one-bitters' for CD players. The one-bit digital-to-analog converter is easy to implement and inherently linear, and the subsequent analog filter can be very simple because most quantization noise is shifted to high frequencies

different implementations and several orders of noise shaping possible [2,21,22,23].

We will settle for a simple first-order configuration as shown in Fig.5-13, and refer to it as a *bitstream generator*, in accordance with its function. The clocked register and the full adder clearly realize the discrete-time integrator function as shown in Fig.5-4. A shift register provides the digital input signal to the other input of the full adder. In that way, the input value can easily be programmed by a serial interface.

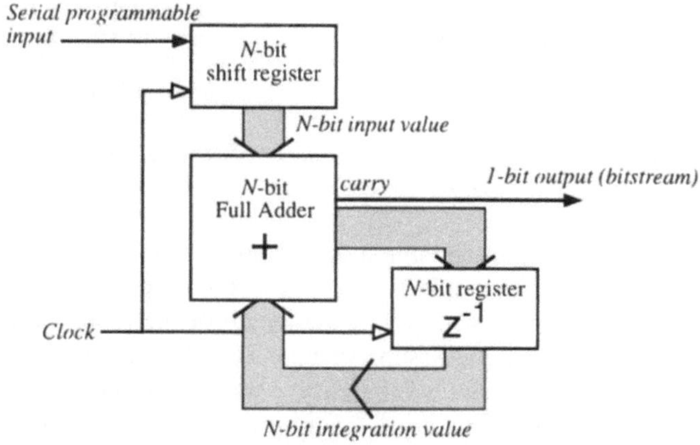

Fig. 5-13 *Configuration of a first-order sigma-delta bitstream generator.*

Because the N-bit full adder outputs a carry signal each time the digital output exceeds the maximum value (2^N-1), it thus functions as a digital comparator with a built-in reference level. The full adder continues its count at a shifted output value as it loses the bits above the most significant bit (MSB). Hence the subtraction of a reference, needed in a sigma-delta converter, is also inherently fulfilled in the digital full adder. The carry signal, which corresponds to a comparator output, is used as the bitstream output signal, proportional to the programmed N-bit input signal. Matters are clarified in Fig.5-14, by showing the signals of a simple 4-bit implementation of the bitstream generator.

One may also notice the difference in the bitstream signal with a pulse-density or pulse-width modulated signal [24]. In fact, the configuration

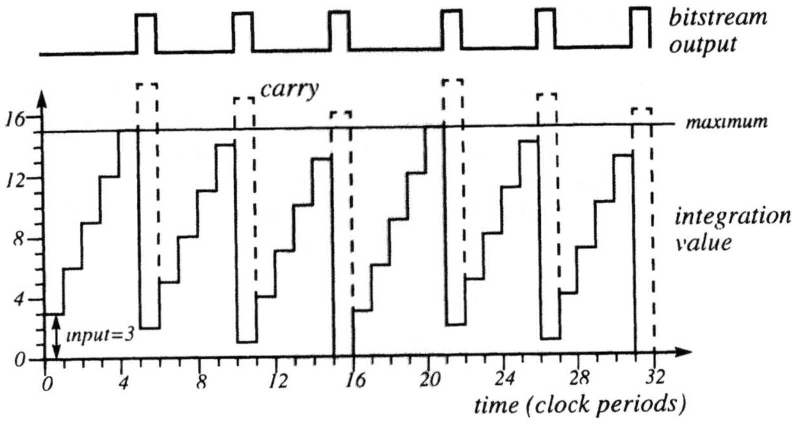

Fig. 5-14 *Digital signal values of the sigma-delta bitstream generator.*

operates as a first-order sigma-delta converter with a (pre-)quantized input signal. Because only quantized input levels are possible, corresponding to a multiple of the fraction 2^{-N}, the length of the produced limit cycles is always limited to 2^N. Assuming a constant digital input signal, the lowest frequency component which may occur in the noise-shaped bitstream output spectrum is the clock frequency divided by 2^N. When filtering the bitstream output at half that frequency or lower, the original digital input signal can be recovered with the same N-bit resolution.

As with the differential sigma-delta AD-converter, we can introduce a polarity to the bitstream signal by assigning a value of -1 (instead of 0) to a low output state. The N-bit input value can be 2's complement coded by inverting the MSB. The maximum symmetrical signal range that can be used goes from $-(2^{N-1}-1)/2^{N-1}$ to $+(2^{N-1}-1)/2^{N-1}$, avoiding the use of a constant (not-switching) bitstream signal. For large N the range practically equals ± 1.

In the following sections we will explain that the generated bitstream signal can be used to modulate the analog input signals of a (normal) sigma-delta AD-converter in order to realize a programmable multiplication factor.

5.3.2 Modulating the input signal for gain calibration

Just like the reference current, the input current may also be switched between the positive and negative input of the integrator, as illustrated in Fig. 5-15. The switch is controlled by a bitstream generator as described in section 5.3.1.

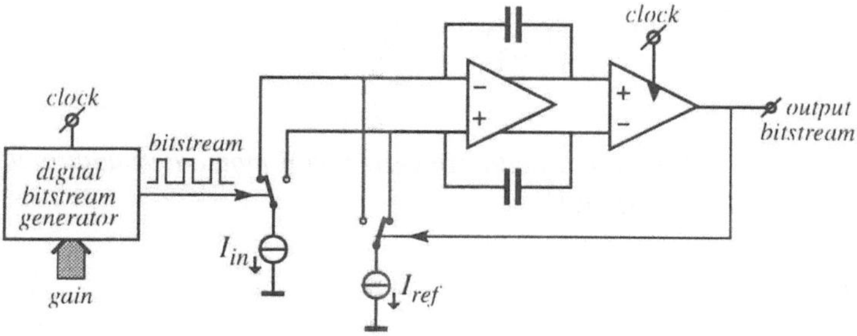

Fig. 5-15 *Sigma-delta converter with modulated input current.*

Imagine, for example, that the input current I_{in} is being switched to the negative side just as many times as to the positive side, the average contribution will then be zero. In fact, the analog input current is simply modulated by the programmable bitstream signal. After averaging or low-pass filtering by a decimation filter, the sigma-delta output will be proportional to the multiplication of the current ratio I_{in}/I_{ref} and the input value of the generator, indicated as gain in the figure.

The average of the generated bitstream, and thus the gain, can be programmed with an accuracy that is determined by the resolution of the bitstream generator. The resolution of this fully digital circuit can be made very high.

In the configuration of Fig. 5-15, the gain may be controlled almost from -1 to +1. Such a large correction range may not be desired. Fig. 5-16 presents an alternative configuration in which a down-scaled copy of the input current is being modulated and added at the integrator inputs.

Provided that the input signal can be duplicated and scaled by a factor α ($\alpha<1$), a gain ranging from $1-\alpha$ to $1+\alpha$ is now possible. The next figure, Fig. 5-17, shows the transfer of the complete sigma-delta converter (with sinc^2 decimation) for different values of the gain. The resolution of the bitstream was 6-bit, α was 0.25, and the window size of the decimation

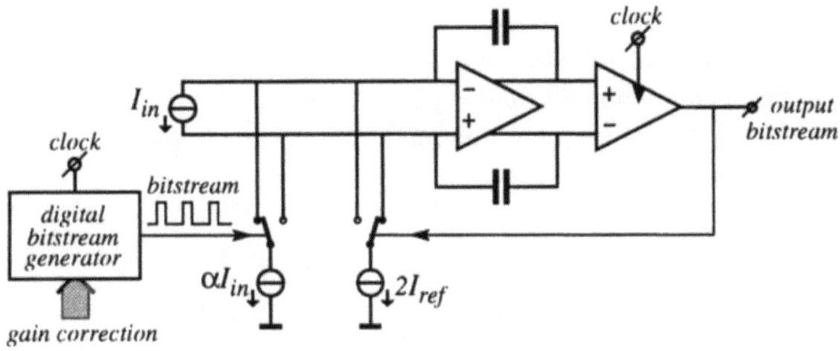

Fig. 5-16 *Sigma-delta converter with modulated additional input signal.*

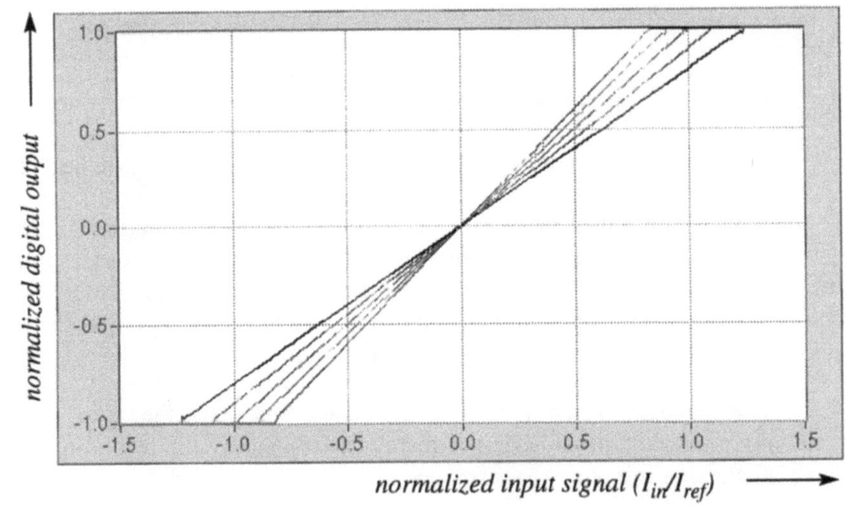

Fig. 5-17 *Signal transfer of the gain-programmable sigma-delta converter, for different gain values.*

filter was 128. For this test case, the digital 2's complement numbers 0, plus and minus $^{13}/_{32}$, and $^{26}/_{32}$ were programmed in the bitstream generator, corresponding to gain values of approximately 0.8, 0.9, 1.0, 1.1, and 1.2.

It is clear that we can map a larger or smaller range of input signals, compared to the usual $\pm I_{ref}$, onto the normalized output range ± 1, so that the correction can be used for a programmable gain calibration of a sensor signal.

To understand the effect on the performance of the AD-conversion and under which conditions the modulation can be used, we examine the behavior in the frequency domain. The first picture in Fig.5-18 shows the frequency spectrum of the bitstream output of a sigma-delta with a normal unmodulated sine-wave input signal, and an oversampling ratio of 100.

The dotted lines in the next picture are the frequency components of the digitally generated bitstream used for the gain correction. Because of the fixed (DC), quantized (digital) input, and the first-order generator, there are only a few discrete spectrum components. The DC-component, not visible because of the logarithmic frequency axis, represents the average bitstream value corresponding to the programmed digital input. The spectrum is shaped in the same way as that of a first-order sigma-delta converter.

The sine-wave input signal that is being modulated by the bitstream results in sideband signals next to each bitstream component, as indicated in the same figure with solid lines. Also, at the signal frequency, a component results which represents the gain correction. Because the amplitude of these sideband components is proportional to that of the modulating frequency (of the bitstream), also the modulated correction signal has a shaped noise spectrum.

The modulated correction signal is added at the input of the sigma-delta converter, and thus some additional shaped 'noise' is introduced. The level of the noise can be lowered by reducing α[1], which controls the amplitude of the correction signal. Furthermore, the presence of this 'noise' effects a dithering of the integrator signal, which results in generation of a much denser noise spectrum. The last picture of Fig.5-18 shows the spectrum of the bitstream generated by the sigma-delta AD-converter with modulated input signal. To demonstrate the effect, a large factor $\alpha=1$ was chosen. It can be seen that extra noise peaks appear, that the shape of the quantization noise has been preserved, and that the amplitude of the input signal has been reduced by the gain calibration.

The effect of the modulation is thus twofold. On one side the performance of the AD-converter is affected in a negative way: an extra noise source is

[1] α can be seen as the *modulation depth* of the input signal

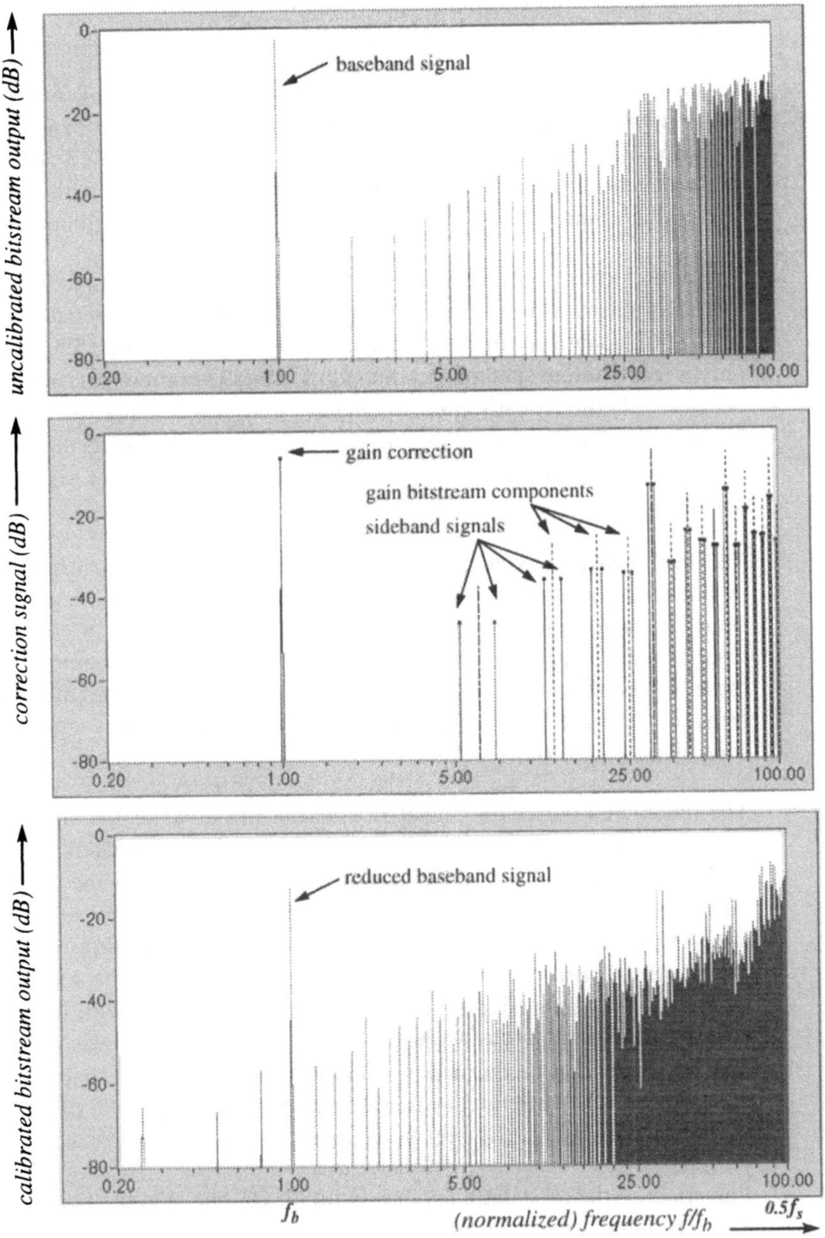

Fig. 5-18 *Spectra of, respectively, the unmodulated bitstream output, the correction signal, and the modulated output.*

introduced at the input. The (frequency-)shape of that noise is determined by the digital sigma-delta, and the amplitude depends on the factor α, which also controls the range of the gain correction. If α is small enough, suppose a 25% gain correction range is sufficient, or if the noise shaping of the correction signal is better than that of the sigma-delta AD-converter, the introduced noise can be neglected.

On the other side, the behavior of the AD-converter is improved due to a dithering effect on the input signal. The noise-shaped output spectrum, which for a first-order usually consists of many discrete peaks, is smoothened to approach the theoretical shape more closely. This improvement especially applies for DC-input signals.

5.3.3 Modulating an additional signal for offset calibration

In the same way as for the gain correction, we may add a modulated reference current at the input of the sigma-delta converter, as shown in Fig.5-19. This allows a programmable offset correction of the transfer of the sigma-delta converter. Also here the correction range can be set by a scale factor α, for the copy of the reference current.

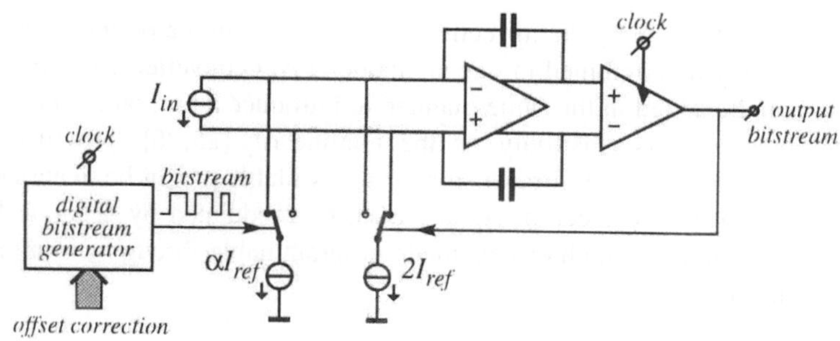

Fig. 5-19 *Sigma-delta converter with modulated additional reference current.*

The effect on the transfer is shown in Fig.5-20, using the same conditions as for the gain correction: DSR=128, α=0.25, and a 6-bit bitstream generator programmed with the digital numbers 0, plus and minus $^{13}/_{32}$, and $^{26}/_{32}$.

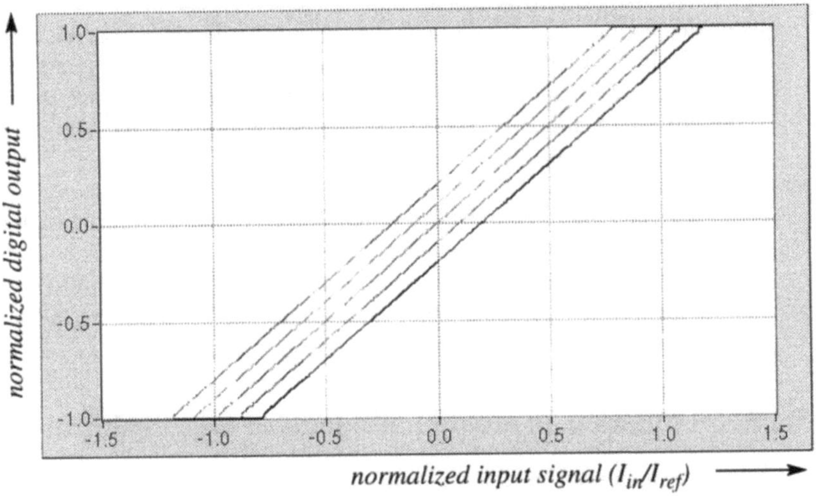

Fig. 5-20 *Signal transfer of the offset-programmable sigma-delta converter, for different offset values.*

5.3.4 Sigma-delta calibration configurations

The offset and gain corrections, as presented in the previous sections can simply be combined in one sigma-delta AD-converter. Even more signals can be added in the same manner, to introduce a programmable correction of the cross-sensitivity or the nonlinearity [25,26], provided that the required analog (current) signals are available or can be generated easily. The analog correction signals must be modulated by separate bitstream generators[1], which can be made programmable through a common serial interface.

As an example, Fig.5-20 shows a sigma-delta configuration with a modulated current $\alpha_1 I_{in}$ for gain control, a modulated current $\alpha_2 I_{ref}$ for offset control, and, additionally, a temperature dependent current I_T for control of the temperature-coefficient in the offset of the transfer. To avoid confusion, it must be mentioned again that the aim of the circuit is to calibrate the overall transfer of a smart sensor, not that of an AD-converter.

[1] Multiple use of a single bitstream generator through multiplexing is also possible, but somewhat complicated, because for the integration value must also be preserved Another option is to multiplex the generator over very long time-slots (oversampling factor). Then the window size of the decimation must be increased accordingly

The input current should be deducted from the sensor signal, the temperature current can be obtained by use of a PTAT source or another temperature sensor.

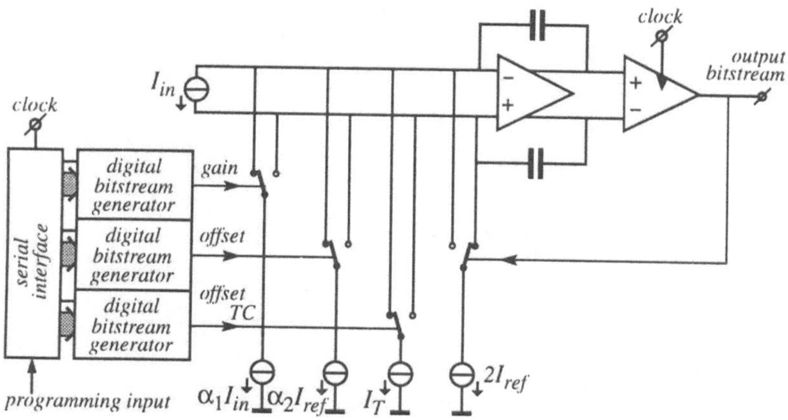

Fig. 5-21 *Sigma-delta configuration with programmable offset, offset temperature-coefficient, and gain.*

To realize a correction possibility for the temperature-coefficient in the gain of the transfer, a signal proportional to the product of I_{in} and I_T is needed. One approach may be to use an analog multiplier, as suggested in section 4.4. However, the multiplication can also be achieved with an additional modulated sigma-delta AD-converter for the temperature signal, as illustrated in Fig. 5-20. First, a bitstream which is proportional to the temperature and a programmable calibration factor is generated. This bitstream is used to modulate an additional current source proportional to the input signal which realizes the multiplication.

Even more configurations become possible when it is conceived that the bitstream signals can also be added in the digital signal domain, as suggested in [28,31]. This would require a separate sigma-delta converter for each analog signal, and a noise-shaping digital adder at the output, in which the bitstreams are weighted and added[1].

[1] The author believes there still lies a research field here to be explored, but already initiated by [28,...,34]

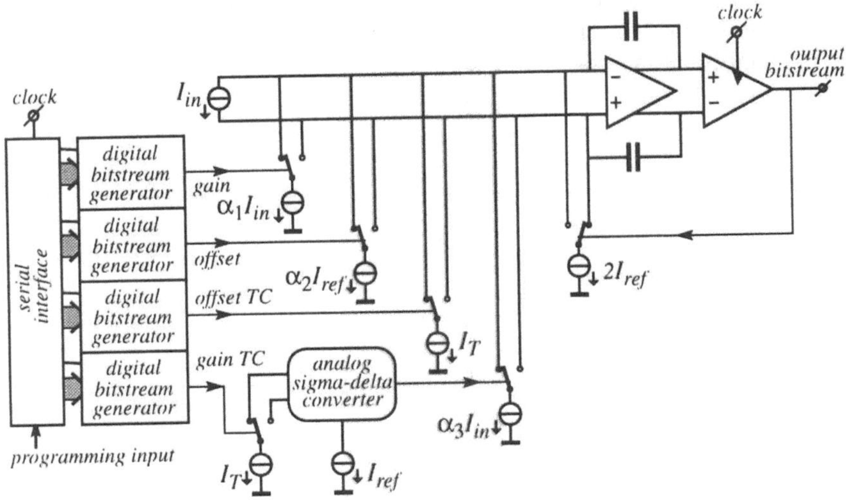

Fig. 5-22 *Sigma-delta configuration with bitstream calibration of offset, gain, and temperature coefficients of both.*

Two things become very apparent in the presented configurations when looking at the many (modulated) input signals that are summed at the input of the integrator.

First, the integrator is loaded with a much larger input current than usual, because at some point in time all the currents may get summed. To avoid clipping of the integrator output, either the integration capacitors must be increased, or the input currents must be scaled down. The first requires a larger bias current of the integrator output stage and demands a larger dynamic range for the comparator input. The latter affects the dynamic range requirements at the integrator input. In both cases more bias current is required to maintain the (analog) signal-to-noise ratio.

Second, the multiple randomized input signal improves the dithering effect, because all the bitstream generators add a sort of pseudo-random noise on top of the DC average correction signals. This especially manifests itself when the bitstream generators produce wide, but shaped noise spectra. This happens for irregular bitstream patterns generated by high-resolution input values ($^{123}/_{512}$, for example) or with higher-order modulators.

The positive effect on the noise shaping of the AD-converter's bitstream output can be seen in the spectrum shown in Fig. 5-23. It resulted from a sigma-delta converter with an arbitrary offset and gain calibration, using modulation depths $\alpha_1=\alpha_2=0.25$. The spectrum shows a good resemblance to the theoretical first-order noise shape.

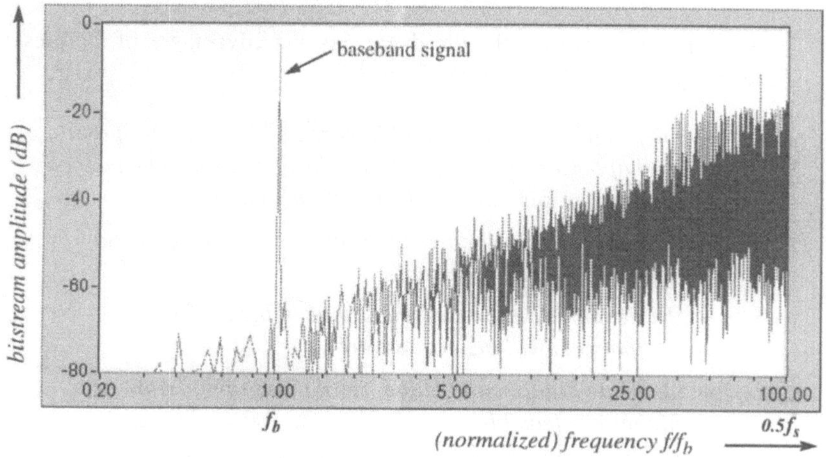

Fig. 5-23 *The bitstream spectrum of a first-order sigma-delta converter with modulated input signals, and OSR=100.*

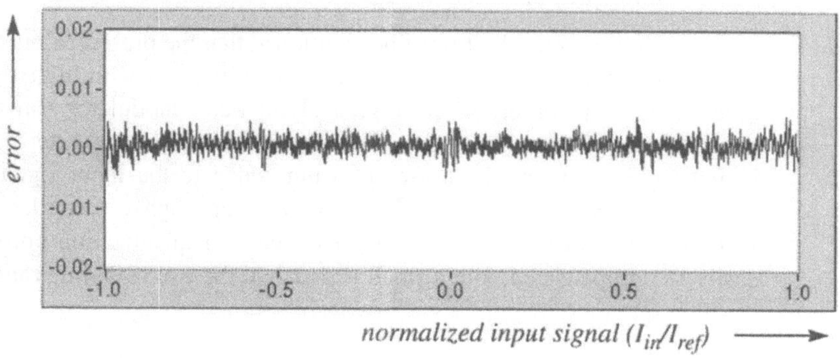

Fig. 5-24 *Error versus input-level for a first-order curve sigma-delta converter with modulated input signals, using a second-order decimation with a window size of 128.*

The positive effect on the error for DC input signals is demonstrated by the error function shown in Fig.5-24. It was obtained by simulating a sigma-delta modulator with a small offset and a small gain calibration, and by using a second-order decimation with a window size of 128. When comparing Fig.5-24 with the original error curve in Fig.5-12, we see that the error has been smoothened, but that the average has gone up slightly.

To summarize, we list the disadvantages and advantages of the presented technique for smart sensor calibration:

- *the progressive polynomial linearization method proposed in chapter 3 cannot easily be applied here, since the method is based on summation of correction signals*
- *preferably, a small correction range should be used, say below 50% of the full-scale*
- *the demands on the dynamic range of the analog components will be more severe*

+ *accurate digital calibration, resolutions up to 16-bit or higher are feasible*
+ *simple, robust, and accurate analog-to-digital conversion which does not rely on device matching*
+ *one-bit serial correction signals are used, thus requiring a minimum number of routing wires from the digital to the analog part of the circuit*

As a concluding remark, it must be mentioned that the presented bitstream modulation technique can also be used in combination with higher-order modulators. The order of the digital sigma-delta modulator (bitstream generator) as well as the order of the sigma-delta AD-converter can be increased. Because of the noise spectrum added to the input signal, the order of the digital sigma-delta should be at least as high as the order of the AD-converter. In that case, the small increase of quantization noise can easily be compensated with a small increase of the oversampling ratio.

5.4 Implementation of the sigma-delta AD-converter

The main analog building blocks of the sigma-delta converter are the differential amplifier and the clocked comparator. The implementation of those blocks will be addressed in the subsequent sections 5.4.1 and 5.4.2.

The current switches have been implemented simply by differential pairs, preceded by emitter followers to obtain a level shift. The circuitry for the generation of the current signals depends on the sensor application. For example, the PTAT and reference current source of section 2.3.1 and the VI-converters of section 4.5.4 can be used. The operation of the presented blocks is demonstrated by a simulation shown in section 5.4.3.

The digital blocks were designed in such a way that implementation is possible, using standard digital CMOS cells or the bipolar ECL cells, developed for the *Dimes01* process. The latter option, although not very compact, allows integration with the bipolar analog part on one chip. Therefore, the ECL logic was used for the bitstream generators.

A second order decimation filter was realized as a test-chip in *DiMOS01*, a 1.6μ CMOS process which only recently became operational in DIMES. It will be issued in section 5.4.4. Besides the hardware implementation of the bitstream decimator, a first-order decimation (counter) in the read-out software on a PC was also available for testing the sigma-delta converter.

5.4.1 Differential integrator amplifier

The differential amplifier is used in the integrator of the sigma-delta converter. The integrator amplifier fixes the voltage level at the input, and reduces negative influence of the Early effects in the currents sources and switching transistors, while providing the integrated large-swing voltage signal at the output.

The amplifier used is shown in Fig.5-25. It is a very simple 2-stage amplifier design, built up with a differential pair input stage Q1-Q4 (quad), a folded cascode, Q7 and Q8, and two voltage followers, Q11 and Q12, for providing the output current. The voltage amplification is realized by the input stage cascode combination only. The dominant pole is at the bases of the output stages. No compensation capacitors were needed so that the amplifier has no internal slew-rate limit. A voltage gain of 66dB (2000) and unity-gain bandwidth of 20MHz were obtained. The current gain was comparable (~1200). These figures are sufficient to neglect integrator leakage for sigma-delta resolutions up to 10-12 bit [4,35].

Furthermore, the amplifier contains two common-mode (CM) control loops, one at the input and one at the output. Transistors Q5 and Q6, in combination with the level-shifted connection through R5 to the common emitter node of the input stage, fix the CM input level to a few hundred

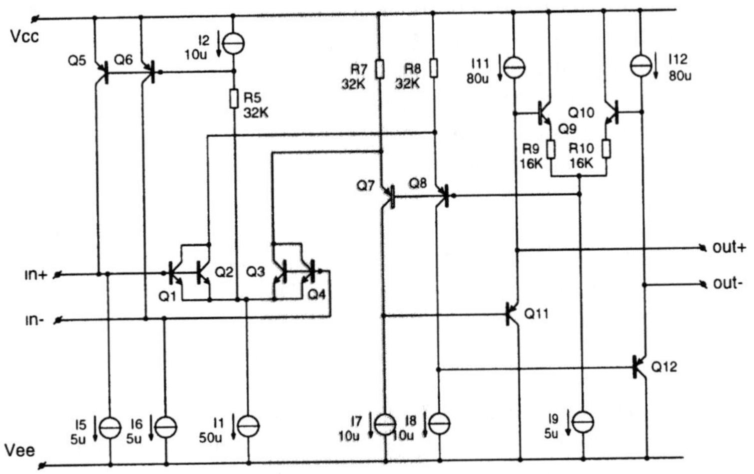

Fig. 5-25 *Differential design of the integrator amplifier.*

millivolts under the supply. This allows to drive the current switches, that are connected to the integrator input, by digital voltage signals on ECL level (see section 4.5.5) without saturating them.

The second CM-control does not really fix the CM-level of the output voltage, but regulates the maximum of both voltages. The degenerated differential pair, Q9 and Q10, detects which of the two output voltages is higher, and the common node (with R9 and R10 connected) follows that voltage. A feedback through the cascode transistors Q7 and Q8 combined with emitter resistors R7 and R8 prevents the output voltage from rising too far. The CM-control 'gently' clips the voltage at the rising output node just below the supply level, while the voltage the other output can go down. The asymmetry of this control with respect to the differential signal, is compensated by the differential feedback configuration in which the amplifier is used. The scope plot shown in Fig.5-26 demonstrates the operation of the control loop. While the individual output voltages are distorted, the differential output voltage amplifies the input correctly.

An advantage of such a control is that when the differential voltage is small, and the connected comparator has to make a precise decision, the absolute voltage levels are close to the supply voltage. This means the (NPN) input stage of the comparator has to be operational only in a small range around the positive supply (not, as usual, halfway the supply rails)[1].

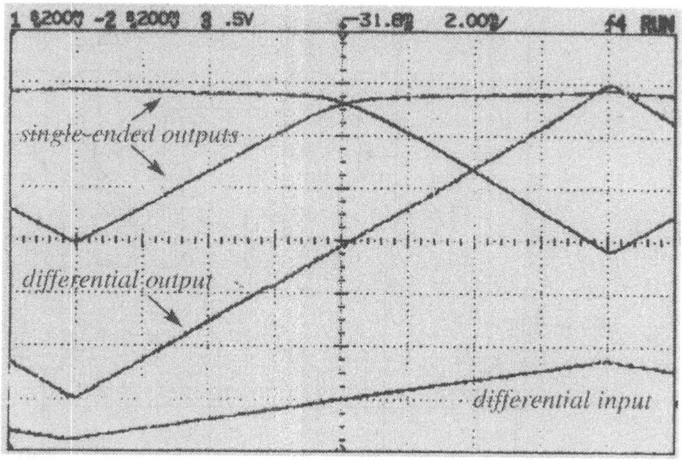

Fig. 5-26 *Input and output signals of the differential amplifier, used in a feedback configuration.*

The presented amplifier design has been completed with the bias circuitry for the current sources and was implemented in a very compact standard layout cell for the analog library. It is shown in Fig. 5-28.

5.4.2 Clocked comparator

The clocked comparator uses the familiar master-slave configuration and is shown in Fig. 5-27. The amplifying stage, Q1 and Q2, and the latching stage, Q3 and Q4, are turned on and off alternately, controlled by the clock signal. The other combination, Q5, Q6 and Q7, Q8, is switched in the opposite way. On positive clock edges, the input voltage amplified onto R1 and R2, is latched by Q3, Q4, and passed on by Q5, Q6 to the output. At the negative clock edge the second latch Q7, Q8 holds the output state.

The input pair is biased with an additional current source I1 to avoid that it is switched off completely. The emitter followers, Q13 and Q14, buffer the clock signal onto the clock switches, so that the digital clock signal can also be provided on ECL level. Saturation of the transistors will not occur, provided that the input signal is provided at a high CM level, as discussed at the end of the previous section.

[1] Especially in low-voltage applications the proposed asymmetrical cm-control will thus be very practical.

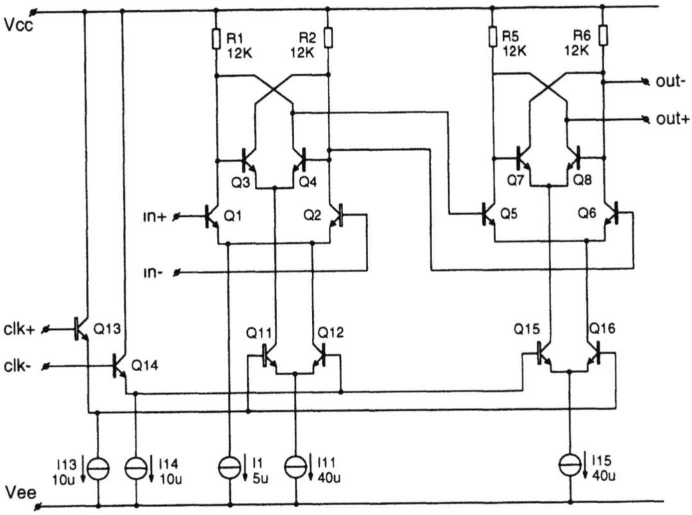

Fig. 5-27 *Differential design of the clocked comparator.*

Also, the comparator has been completed with the transistors for the current sources, and laid out in a compact standard cell. It was added to the analog cell library, which was also used for the design in chapter 4. Fig. 5-28 gives an impression of the layout cells of the two function blocks discussed here.

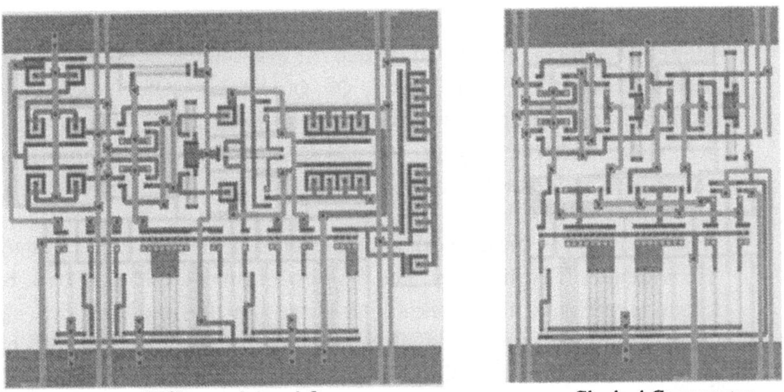

Differential Amplifier Clocked Comparator

Fig. 5-28 *Standard layout cells of the differential amplifier and the clocked comparator.*

5.4.3 Sigma-delta converter

The functionality of the blocks has been tested by means of a PSPICE simulation of the differential sigma-delta configuration. Depending on the clock frequency the value of the input currents and the supply voltage, smaller or larger integration capacitors must be used. Fig. 5-29 shows the result of a simulation for a clock frequency of 200 kHz, a reference current of 12 μA, and integration capacitors of 40 pF. The supply voltage was 3 Volt. After start-up the integrator signal clearly shows the desired behavior for a first-order sigma-delta modulator.

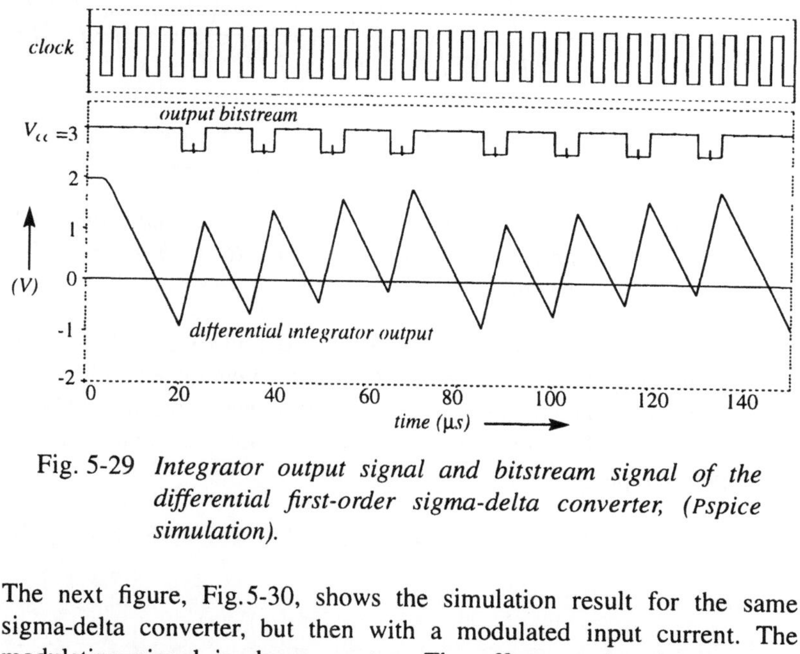

Fig. 5-29 *Integrator output signal and bitstream signal of the differential first-order sigma-delta converter, (PSPICE simulation).*

The next figure, Fig. 5-30, shows the simulation result for the same sigma-delta converter, but then with a modulated input current. The modulating signal is shown on top. The effect on the slopes of the integrator signal is clear: there are now 4 possible input conditions. The maximum voltage-swing also increases.

The computation effort for the PSPICE simulations is too much to simulate enough clock periods for determination of the resolution on the basis of the bitstream output signal.

Using the laid-out analog library cells presented in the previous sections, completed with layout cells for a PTAT bias source, for the current switches

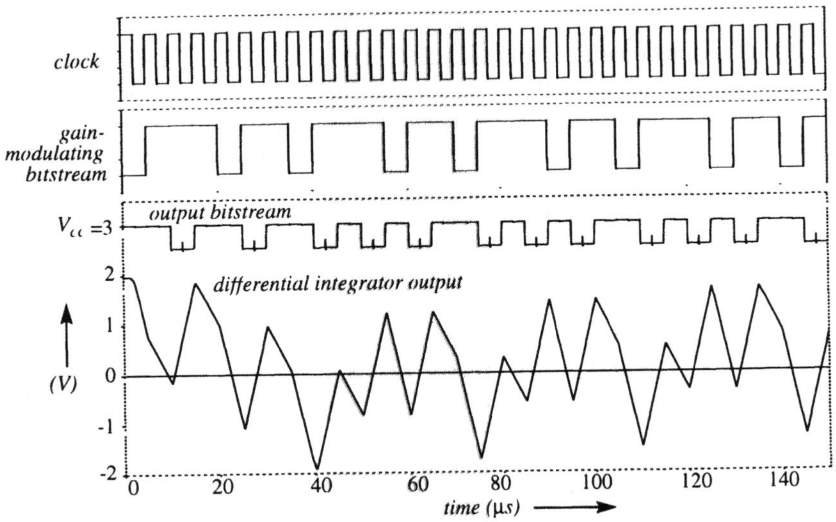

Fig. 5-30 *Integrator output signal and bitstream signal of the first-order sigma-delta converter with modulated input signal.*

and for the capacitors, the layout for different sigma-delta configurations could be generated automatically. It can be combined with automatically generated digital designs for the bitstream generators, as will be shown in section 5.5. Fig.5-31 shows one of the sigma-delta converters on a chip photograph.

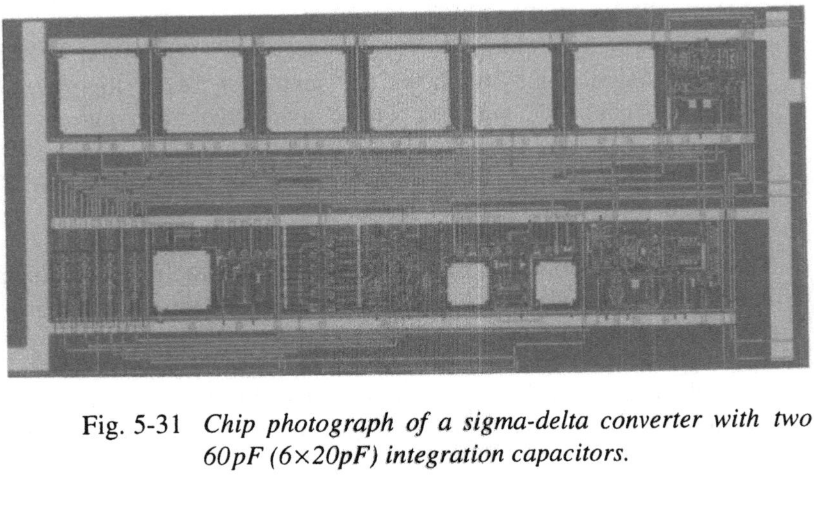

Fig. 5-31 *Chip photograph of a sigma-delta converter with two 60pF (6×20pF) integration capacitors.*

5.4.4 Digital decimation filter implementation

As a test chip for the new DiMOS process, a bitstream decimation filter with smart sensor bus interface has been developed. Fig. 5-32 shows the configuration.

After reset, an 8-bit up/down counter generates the triangular shaped weighting function for a window size of 512 clock periods. Each clock period, the weighting factor from the up/down counter will be added to the memorized sum if the bitstream input is high. When the count returns to zero, at the 512th clock period, the total sum will be stored in the output register. The maximum output equals 512×256/2=65386, hence the 16-bit resolution for the full-adder and the register in the loop. The decimation result can be shifted out of the output register serially, which makes it easy to pass it through to the bus interface. It also makes it possible to process the bitstream and read out the converted signal at different clock frequencies.

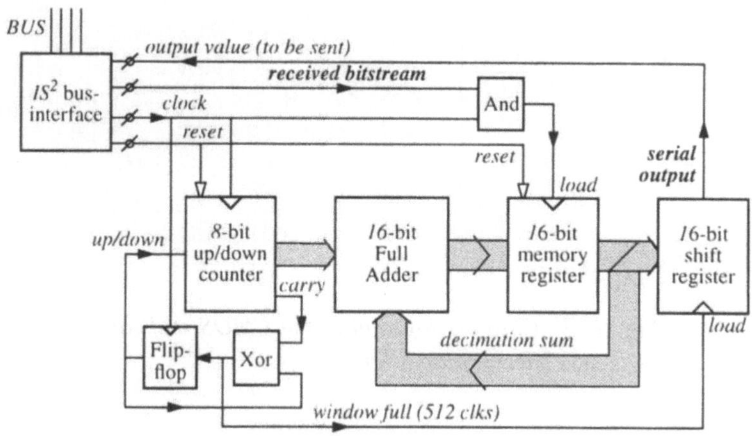

Fig. 5-32 *Configuration of the 16-bit sinc^2 decimation filter.*

The bus interface is connected in such a way that any bitstream put on the bus by a sensor will be decimated by the filter. The bus master must first address a sensor which outputs a bitstream, then, after at least 512 clock periods, it can address the filter to read out the decimated value.

A digital standard cell library was adapted to the DiMOS design rules, so that the layout could be generated by the *place & route* routine of the layout program L-Edit. Fig.5-33 shows a photograph of the resulting CMOS chip, which measures approximately 3×2 mm^2.

The chip can be used to connect any sigma-delta output to the integrated smart sensor bus, for easy read-out and testing on a PC-controlled measurement set-up. It has been tested, using the integrated smart sensor (IS^2) bus and several smart temperature sensors, as presented in the next section.

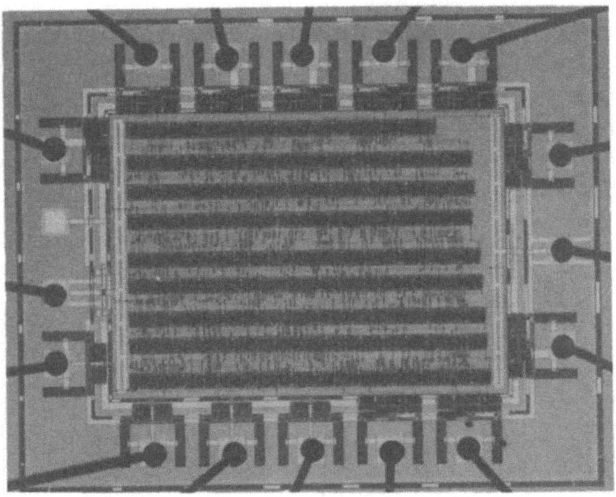

Fig. 5-33 *Chip photograph of the CMOS realization of the 16-bit decimation filter.*

5.5 Smart temperature sensor realizations

The techniques presented in section 2.3.1 and section 5.3 can be combined in the design of a smart temperature sensor with programmable offset and gain calibration. The goal of the calibration is to overcome the device-to-device variations by means of calibration, so that all sensors can then be delivered with a fixed transfer from temperature to output. It is also possible to allow a customer programmable (variable) temperature range. These possibilities have been worked out in the next sections.

5.5.1 Smart temperature sensor with calibration and IS^2 bus interface

Fig. 5-34 shows the configuration of a smart temperature sensor based on the combination of (1) a PTAT and bandgap reference circuit, (2) a sigma-delta AD-converter, (3) an IS^2 bus interface, and (4) two programmable bitstream generators. The combination of a PTAT temperature sensor and a sigma-delta AD-converter has been reported earlier [36].

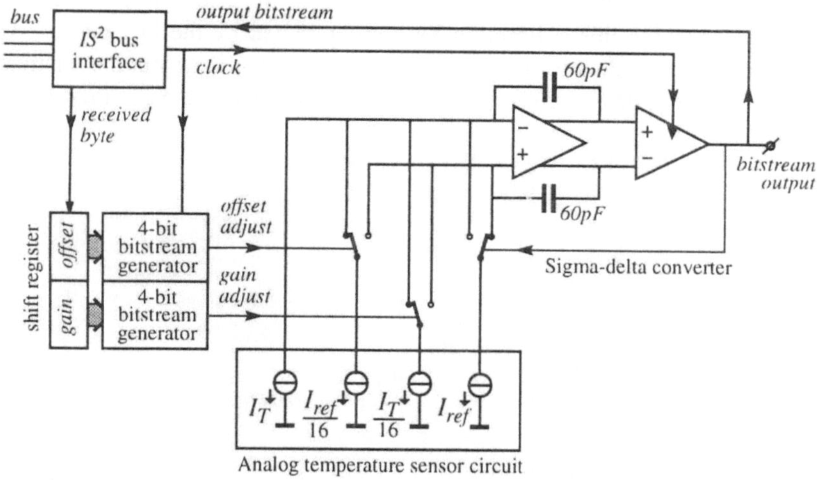

Fig. 5-34 *Configuration of the smart temperature sensor with bus interface and programmable calibration.*

The analog circuit for the temperature sensor is constructed in such a way that a temperature signal I_T and a reference signal I_{ref} are generated, as well as copies of both currents which are scaled down by a factor of 16. It is aimed to obtain a ratiometric signal I_T/I_{ref} which varies from 0 to 1 for the temperature range of -30°C to +70°C. The ratiometric conversion is obtained by the sigma-delta converter. Relatively large integration capacitors 60pF are used, because the maximum sigma-delta clock frequency that currently can be obtained on the smart sensor bus with a PC bus master is low, about 70kHz.

The signal $I_{ref}/16$ is used to correct the offset error in the transfer. It can be added or subtracted from the normal input signal with a 4-bit precision, using the first bitstream generator. In the same way, the signal $I_T/16$ is modulated by the output of the second bitstream modulator, and added at the sigma-delta input for correction of gain errors. The maximum correction for offset and gain is thus approximately ±6.25% of the full range. This is sufficient to compensate for the errors that can be expected to result from process variations, provided that the process is characterized well so that the actual nominal (average) sensor transfer is close to the desired transfer (see section 3.1.2).

The two 4-bit digital values for the offset and gain can be programmed through the bus interface. At one of the two sensor addresses the sensor can receive this calibration byte programmed by the bus master. At the other bus address of the sensor, the bus interface sends out the bitstream of the sigma-delta converter which can be received by the bus master.

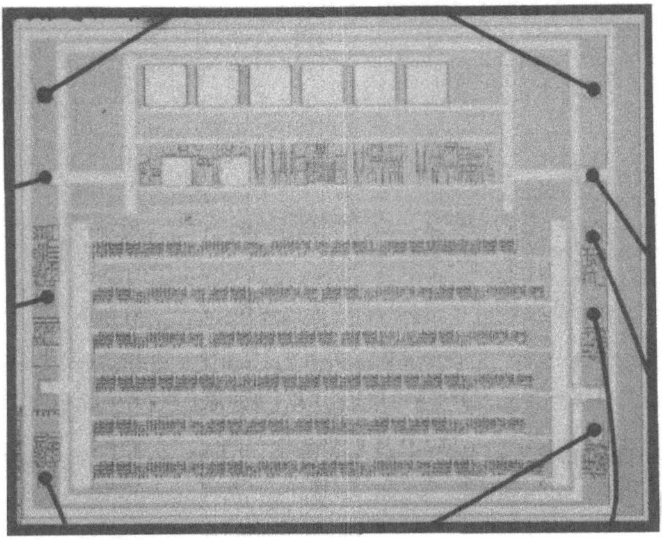

Fig. 5-35 *Chip photograph of the bipolar realization of the smart temperature sensor.*

Fig.5-35 shows a photograph of the chip which was realized in the Dimes-01 bipolar process of the research institute DIMES. It measures approximately 4.4×5.3mm^2. A commercially more feasible chip size may be obtained by integration in a high-density bipolar or in a CMOS process (CMOS temperature sensors have been reported in [37,38], for example) which would reduce the size of the digital part especially. In addition, the size of the analog part may be reduced by shrinking the integration capacitors, and running the sigma-delta converter at a higher clock frequency. In all, the chip size can be reduced to below $1\times1\,mm^2$.

Through the use of the smart sensor bus, several sensors can be measured at the same time, including a reference temperature sensor for calibration purposes. In addition, the decimation filter presented in section 5.4.4 can be connected to the bus for decimation of the bitstream output of each addressed temperature sensor. An impression of the measurement set-up is given in Fig. 5-36. Multiple temperature sensors can be connected to the bus and inserted together in a temperature chamber (oven) for collective calibration.

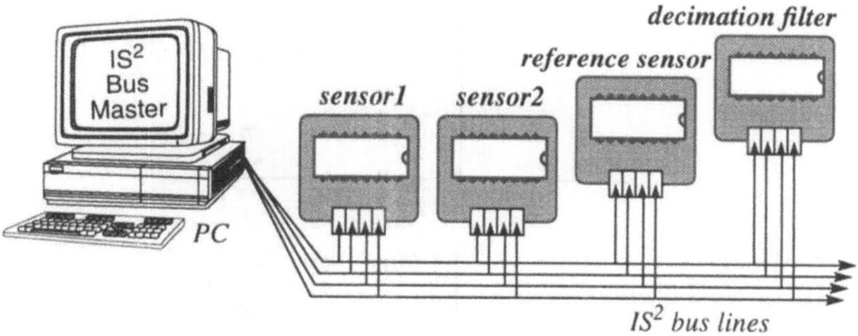

Fig. 5-36 *Impression of the measurement set-up using the integrated smart sensor bus for communication.*

Fig. 5-37 presents the transfer and error of two measured temperature sensors before and after calibration of offset and gain.

Clearly, the curves can be calibrated onto the desired transfer curve. After calibration, the error remains almost within the ±0.5% range, corresponding to an 8-bit resolution. The calibration step size (LSB) was 0.8%, however a noise error was present as well. The parabolic error curve, remaining after offset and gain calibration, can be recognized with difficulty.

The table below summarizes the specifications of the chip. Because of the ECL realization, the digital part consumes a considerable amount of supply current which explains the self-heating of about 0.5 °C.

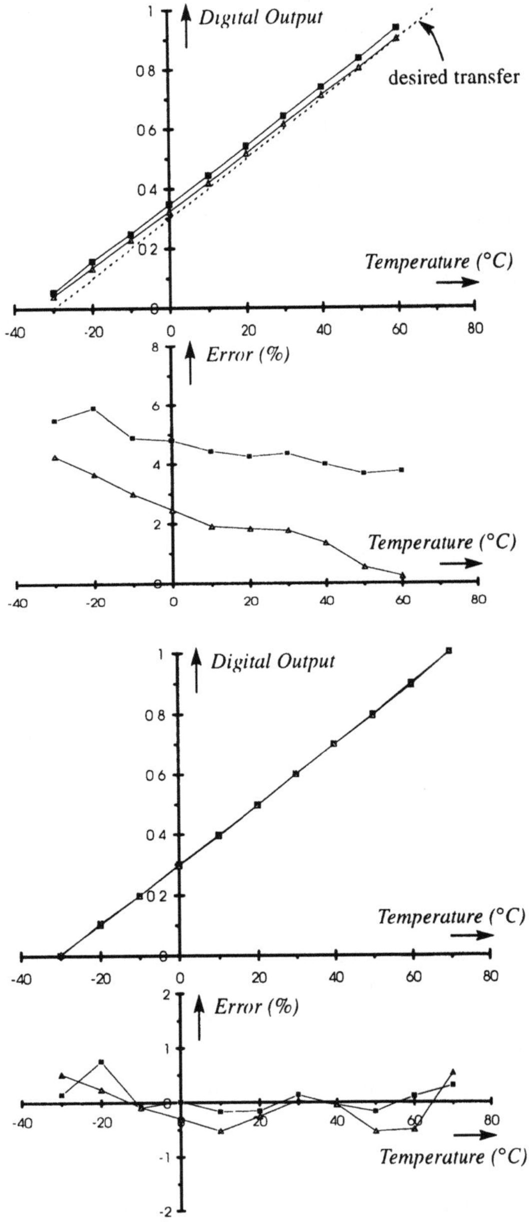

Fig. 5-37 *Measured temperature transfer and relative error of two smart temperature sensors, before and after calibration.*

Table 5-1 *Specifications of the smart temperature sensor with gain and offset calibration, and bus interface.*

Description	Value	Unit
Temperature range	-30 to +70	°C
Calibration accuracy (LSB)	0.8	% of FS
Calibration range, offset & gain, 4-bit	±6.25	% of FS
Useful output resolution	7-8	bit
Maximum error after calibration	±0.8	% of FS
Clock frequency	50-200	kHz
Supply Voltage, Vcc	3-6	V
Supply current, analog part	0 4	mA
Supply current, digital part	1 6	mA
Self-heating	~0.5	°C

Conditions· Vcc=5, DIL16 package

5.5.2 Temperature sensor with linearity calibration

For a more accurate temperature sensor, a correction of the nonlinearity is needed. When using the ratio of a PTAT signal and a bandgap reference, the error typically shows a parabolic curve [36,...,39]. It can be shown that this is in fact a logarithmic error, of which the Taylor polynomial has a dominant quadratic component [39]. This can be compensated with a signal based on the product of a PTAT signal with a positive temperature coefficient, and a V_{be} signal which has a negative temperature coefficient. The multiplication can be obtained with a simple analog current multiplier, as shown in Fig.5-38.

The PTAT, V_{BE}, and reference signals are derived from a temperature sensor signal as presented in section 2.3.1.

The temperature signals can be combined with the bitstream calibration technique, as proposed in the configuration of Fig.5-39. This smart temperature sensor has been presented in [27].

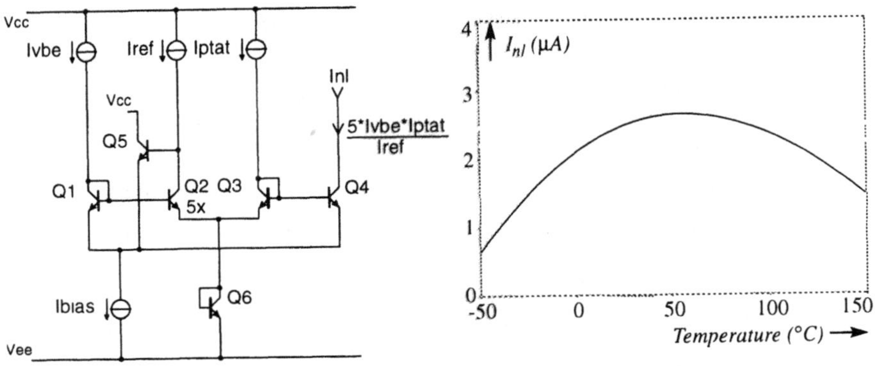

Fig. 5-38 *Current multiplier for generation of a parabolic compensation signal, and output current I_{nl} versus temperature.*

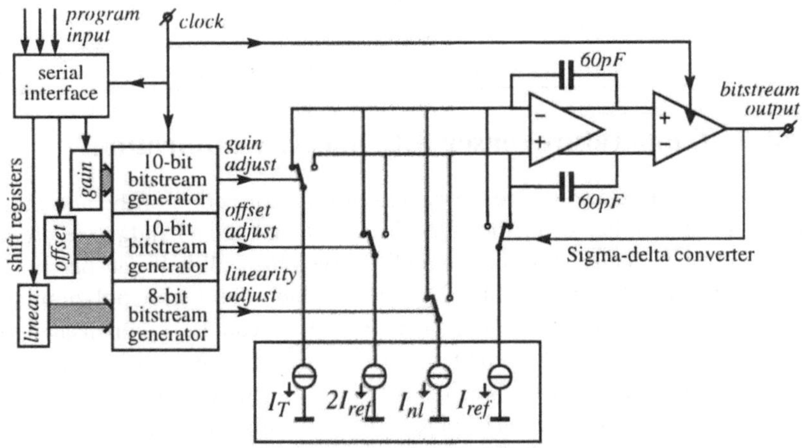

Fig. 5-39 *Configuration of the second, programmable smart temperature sensor.*

Using high-resolution bitstream generators the three different current signals I_{ref}, I_T, and I_{nl} can be accurately weighted with a digital value, and summed at the input of the sigma-delta converter. The sum is converted with respect to the sigma-delta reference current, into a digital output. The three ratiometric signals $2I_{ref}/I_{ref}$, I_T/I_{ref}, and I_{nl}/I_{ref} have respectively zero-order (constant), first-order (linear), and second-order (quadratic) dependency on the temperature, as shown in Fig. 5-40. The supposedly linear signal, however, has a small quadratic error. The contribution of each of these signals can be varied from -1 to +1, so we have full control of the overall temperature transfer. Through a serial interface the bitstream generators can be programmed in such a way that the temperature can be measured accurately over a certain range.

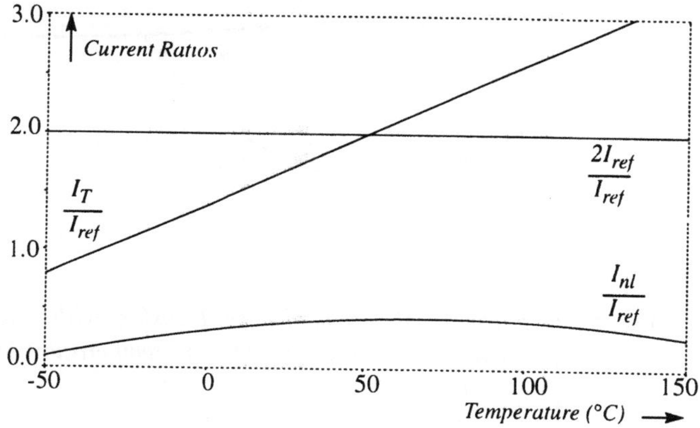

Fig. 5-40 *The three ratiometric temperature signals used in the second smart temperature sensor.*

A calibration should proceed as follows. At three different temperatures, preferably 'spaced' as far apart as possible, each of three programmable signals must be measured separately. This can be done by consecutively programming the bitstream generator of each signal with a fixed value, and the other two with zero. The three signals can then be characterized precisely, using second-order polynomials for the two temperature-dependent signals. For a desired linear temperature transfer it can then be calculated what the contributions of each of the three signals should be. The required values can then be programmed in the shift registers of the chip. The sigma-delta output range can be mapped on a temperature range of at least 75 °C, and in between the operation limits of the IC, from -40°C to 130°C.

Fig. 5-40 shows how the parabolic linearity error, remaining after offset and gain calibration, can be reduced to a much smaller third-order linearity error, using the addition of a quadratic temperature signal.

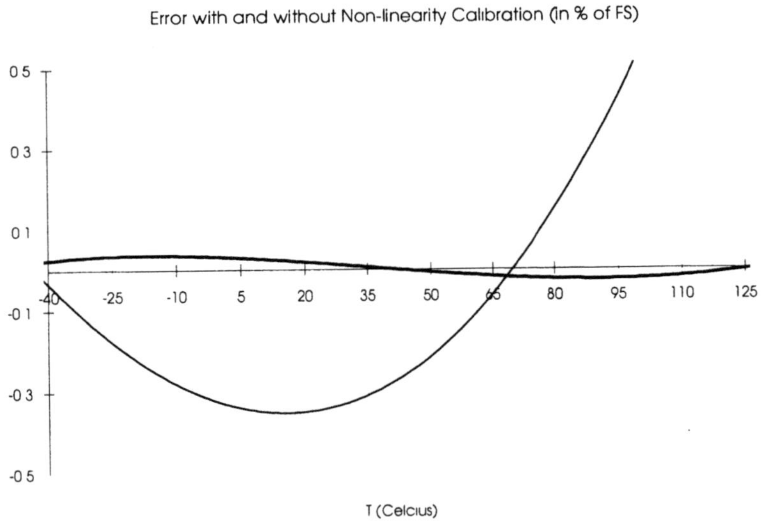

Fig. 5-41 *Error in the temperature transfer, with and without linearity calibration, for a temperature range of -40 to 125°C.*

5.6 Conclusion

The construction of the sigma-delta Analog-to-Digital converter allows modification of the transfer by summing several analog input signals which are modulated by digitally generated bitstream signals. The sigma-delta (noise-shape) techniques can also be applied in the digital bitstream generation, which makes it possible to calibrate very accurately. However, the suggested configurations only allow for a low order of calibration.

The bitstream technique is appropriate for integration in the smart sensor concept, as demonstrated with the implementation of a smart temperature sensor with bus-interface and programmable calibration in a bipolar chip. For high accuracies, the required digital circuitry can only be integrated in a feasible manner, using a more compact CMOS technology.

References

[1] J.C. Candy and G.C. Temes, *Oversampled Delta-Sigma Data Converters*, IEEE Press, New York, 1992.

[2] R. v.d. Plassche, *Integrated Analog-to-Digital and Digital-to-Analog Converters*, Kluwer Academic Publishers, Boston/Dordrecht/ London, 1994.

[3] B.E. Boser, *Design and implementation of oversampled analog-to-digital converters*, PhD Thesis, Stanford University, 1989.

[4] F.R. Riedijk, *Integrated Smart Sensors with Digital Bus Interface*, PhD Thesis, Delft University Press, Delft, 1993.

[5] W. Bennet, "Spectra of quantized signals", Bell Syst. Tech. J., Vol.BSTJ-27, July 1948, pp.446-472.

[6] J.C. Candy and O.J. Benjamin, "The structure of quantization noise from sigma-delta modulation", *IEEE Trans. Communication*, Vol.29, No.9, September 1981, pp.1316-1323.

[7] R.M. Gray, "Oversampled sigma-delta modulation", *IEEE Trans. Communication*, Vol.35, 1987, pp.481-489.

[8] R.M. Gray, "Quantization noise spectra", *IEEE Transactions on Information Theory*, Vol.36, No.6, 1990, pp.1220-1244.

[9] J.C. Candy, "A use of double integration in sigma-delta modulators", *IEEE Trans. Communication*, Vol. COM-33, March 1985, pp.249-258.

[10] R.W. Adams, "Design and implementation of an audio 18-bit Analog-to-Digital converter using oversampling techniques", *J. Audio Eng. Soc.*, Vol.34, No.3, March 1986, pp.153-166.

[11] J. Robert, G.C. Temes, V. Valencic, R. Dessoulavy, and P. Deval, "A 16-bit low-voltage CMOS A/D converter, *IEEE Journal of Solid-State Circuits*, Vol. SC-22, April 1987, pp.157-163.

[12] J. Robert and P. Deval, " A second-order high-resolution incremental A/D converter with offset and charge-injection compensation, *IEEE Journal of Solid-State Circuits*, Vol.23, No.3, June1988, pp.736-741.

[13] Y. Matsuya K. Uchimura, A. Iwata, T. Kobayashi, M. Ishikawa, and T. Yoshitome, "A 16-bit oversampling A-to-D conversion

technology using triple integration noise-shaping", *IEEE Journal of Solid-State Circuits*, Vol. SC-22, 1987, pp.921-929.

[14] B.E. Boser and B.A. Wooley, "The design of sigma-delta modulation Analog-to-Digital converters", *IEEE Journal of Solid-State Circuits*, Vol.23, 1988, pp.1298-1308.

[15] J.C. Candy, Y.C. Ching, and D.S. Alexander, "Using triangularly weighted interpolation to get 13-bit PCM from a sigma-delta Modulator", *IEEE Trans. Communication*, Vol.24, 1976, pp.1268-1275.

[16] R.E. Crochiere and L.R. Rabiner, "Interpolation and decimation of digital signals - A tutorial review", *Proceedings of the IEEE*, Vol.69, 1981, pp.300-331.

[17] E. Dijkstra, M. Degrauwe, J. Rijmenants, and O. Nys, "A design methodology for decimation filters in sigma-delta A/D converters", *Proceedings ISCAS'87*, 1987, pp.479-482.

[18] A.V. Oppenheim and R.W. Schafer, *Discrete-time Signal Procesing*, Prentice-Hall, New Jersey, 1989.

[19] L.R. Carley, "An oversampling Analog-to-Digital converter topology for high-resolution signal acquisition systems", *IEEE Trans. Circuits and Systems*, Vol.34, 1987, pp.83-90.

[20] P.C. de Jong, G.C.M. Meijer, and A.H.M. van Roermund, "A new dithering method for sigma-delta modulators", *J. Integrated Analog Circuits and Signal Processing*, Vol.10, August 1996, pp.193-204.

[21] J.C. Candy and A. Huynh, "Double interpolation for Digital-to-Analog conversion", *IEEE Trans. Communication*, Vol.34, 1986, pp.77-81.

[22] L.R. Carley and J. Kenney, "A 16-bit 4'th order noise-shaping D/A converter", *Proceedings of the Custom Integrated Circuits Conference (CICC)*, 1988, pp.21.7.1-21.7.4.

[23] P.J.A. Naus, E.C. Dijkmans, E.F. Stikvoort, A.J. McKNight, D.J. Holland, and W. Bradinal, "A CMOS stereo 16-bit D/A converter for digital audio", *IEEE Journal of Solid-State Circuits*, Vol.22, 1987, pp.390-395.

[24] A.B. Carlson, *Communication Systems: an Introduction to Signals and Noise in Electrical Communication*, McGraw-Hill Book Co., Singapore, 1986.

[25] G. v.d. Horn and J.H. Huijsing, "Sensor calibration using sigma-delta bitstream modulation", *Proceedings EuroSensors VIII Conference*, Toulouse, France, 1994, pp.34.

[26] G. v.d. Horn and J.H. Huijsing, "Sensor calibration using sigma-delta bitstream modulation", *Proceedings National Sensor Technology Conference, 1994*, pp.183-186.

[27] G. v.d. Horn and J.H. Huijsing, "Smart temperature sensor with integrated bitstream calibration", *Proceedings ESSCIRC'95*, Lille, France, 1995, pp.234-237.

[28] N. Kouvaras, "Delta-modulation/P.C.M converter", *Electronics Letters*, Vol.14, No.20, 1978, pp.660-662.

[29] N. Kouvaras, "Some novel elements for delta-modulated signal processing", *The Radio and Electronic Engineer*, Vol.51, No.5, 1981, pp.241-249.

[30] M. Freedman and D.G. Zrilic, "Nonlinear arithmic operations on the delta-sigma pulse stream", *Signal Processing, Elsevier*, Vol.21, 1990, pp.25-35.

[31] P. O'Leary and F. Maloberti, "A bitstream adder for oversampling coded data", *Electronics Letters*, Vol.26, No.20, 1990, pp.1708-1709.

[32] F. Maloberti, "Non conventional signal processing by the use of sigma delta technique: a tutorial introduction", *Proc. IEEE International Symposium on Circuits and Systems*, San Diego, USA, 1992, pp.2645-2648.

[33] F. Op 't Eynde, "A power metering ASIC with a sigma-delta based multiplying ADC", *Digest of Techn. Papers ISSCC*, 1994, pp.186-187.

[34] V.F. Dias, "Signal processing in the sigma-delta domain", *Microelectronics Journal*, Vol.26, No.6, 1995, pp.543-562.

[35] O. Feely and L.O. Chua, "The effect of integrator leak in Sigma-Delta Modulation", *IEEE Trans. Circuits and Systems*, Vol.38, No.11, November 1991, pp.1293-1306.

[36] F.R. Riedijk and J.H. Huijsing, "An integrated absolute temperature sensor with sigma-delta A-D conversion", *Sensors and Actuators A*, Vol.24, 1992, pp.249-256.

[37] P. Krummenacher and H. Oguey, "Smart temperature sensor in CMOS technology", *Sensors and Actuators A*, Vol.21-23, 1990, pp.636-638.

[38] A. Bakker and J.H. Huijsing, "Micropower CMOS temperature sensor with digital output", *IEEE Journal of Solid-State Circuits*, Vol.31, No.7, July 1996, pp.933-937.

[39] G.C.M. Meijer and A.W. van Herwaarden, *Thermal Sensors*, Institute of Physics Publishing, Bristol/Philadelphia, 1994.

Calibration using Digital Signal Processing

6

6.1 Introduction

It is not the aim of this chapter to investigate or discuss all the possible digital forms of sensor calibration techniques, but rather to demonstrate the feasibility of the progressive polynomial calibration method proposed in *chapter 3*.

It is acknowledged that digital compensation is the most attractive technique for high-resolution sensor calibration. The implementation in software for a digital processor enables an advanced, complex but also flexible correction of sensor signals. Hence, for most linearization techniques explained in *chapter 3*, examples of computer software implementations can be found in literature [1,2,3,4]. For the simpler linearization techniques, implementation in microcontroller programs has also been attempted [5,6].

On the other side, a hardware implementation allows to optimize for speed, chip area, or power consumption, but lacks flexibility [7,8]. In the next section we will first shortly consider the possibility to implement the proposed polynomial calibration method in digital hardware.

Using modern computers, it obviously is possible to realize an advanced digital sensor calibration with high orders of linearization and cross-sensitivity compensation. Such might be interesting for instrumentation systems, but not for the integration on a smart sensor. We must recall our purpose: to produce smart sensors of which the output signal is standardized with respect to a reference signal, and to achieve this at minimum production cost.

When a digital (either dedicated or standard) processor can be (hybridly) integrated at an acceptable price, still another cost aspect must be considered. Namely the time required per calibration per sensor. This time can be minimized not only by using an effective calibration method, but also by minimizing the data communication needed between the sensor(s) and the calibration controller (computer). Such can be obtained if the sensor calibration circuit or program is capable of calculating it's own calibration coefficients. In that case, the calibration controller only has to send a signal to indicate which reference signal is applied during the calibration phase, see section 1.4.2. The signal can be sent to many sensors at a time, which then process their own output signal and calculate their own calibration coefficient, all in parallel. It is clear this may save time with respect to the situation in which a central computer has to read each sensor output, calculate each calibration coefficient, and store it in each sensor, one at a time.

For the proposed polynomial calibration method the calculation of the calibration coefficients is not complicated, and may very well be incorporated in the digital implementation.

6.2 Hardware implementation

For clarity the equations for the one-dimensional polynomial calibration procedure, explained in section 3.3.1, are summarized in Table 6-1. Since we now deal with digital numbers and not with physical signals, the reference signal y_{ref} has been left out. It means a different 'unit' is incorporated in each calibration coefficient.

It was already clear from the *cascaded* implementation shown in Fig. 3-16 on page 82, that the procedure has a repetitive character. The same algorithm is repeated for each calibration step, using the previous output $h_{n-1}(x)$ and the polynomial product $\Pi\{h_i(x)-y_i\}$ as inputs. While in analog

Table 6-1 *Overview of equations for the one-dimensional progressive polynomial calibration technique.*

	calibrated function	calibration coefficient
initial	$y = f(x)$	--
step 1 $h_1(x)$	$f(x) + a_1$	$a_1 = y_1 - f(x_1)$
step 2 $h_2(x)$	$h_1(x) + a_2 \cdot \{h_1(x) - y_1\}$	$a_2 = \dfrac{\{y_2 - h_1(x_2)\}}{\{h_1(x_2) - y_1\}}$
step 3 $h_3(x)$	$h_2(x) + a_3 \cdot \{h_1(x) - y_1\} \cdot \{h_2(x) - y_2\}$	$a_3 = \dfrac{\{y_3 - h_2(x_3)\}}{\{h_1(x) - y_1\} \cdot \{h_2(x) - y_2\}}$
	...	...
step *n* $h_n(x)$	$h_{n-1}(x) + a_n \cdot \prod_{i=1}^{n-1} \{h_i(x) - y_i\}$	$a_n = \dfrac{y_n - h_{n-1}(x_n)}{\prod_{i=1}^{n-1} \{h_i(x_n) - y_i\}}$

hardware the multiplication and summing actions can be realized very compactly, in digital hardware such arithmetic functions require complex circuitry. Therefore, it makes more sense to include the algorithm in a *loop* as proposed in Fig.6-1, and so reduce the number of multipliers and adders. The highlighted algorithm is the same as used in Fig.3-16, besides the absence of y_{ref}. The signals indicated in Fig.6-1 are for the n-th and last calibration step, supposing the calibration points y_1 to y_n are known, and the coefficients a_1 to a_n have already been calculated and stored.

The output and correction term of a calibration step are stored in 'sample & hold' registers, so that they can be used as inputs in the next calibration step. For each calibration step i the values of y_{i-1} and the required coefficient a_i are multiplexed from a memory, and the loop is executed once. The first calibration step, for offset, may be simplified and executed first, outside the loop as shown in the figure. At the second calibration step the inputs are connected in such a way that the first multiplier outputs the signal $(h_1(x)$-$y_1)$ which is used for the gain correction. At succeeding

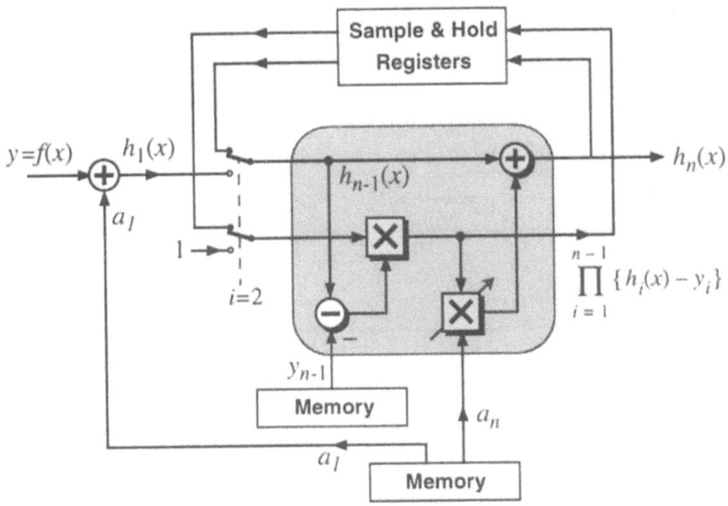

Fig. 6-1 *Algorithmic loop implementation of the one-dimensional polynomial calibration method, at the n-th correction step of the measurement phase.*

calibration steps the inputs are switched to the previous output values stored in the SH registers, and so the product term for the polynomial corrections is built-up.

As mentioned in the introduction, it is desirable that the calculation of the coefficients, during the calibration phase, is incorporated in the implementation itself, and not externally. Knowledge about the desired output values y_1 to y_n can be built in or received from the calibration controller. By expanding the block diagram as illustrated in Fig.6-2, the calibration coefficient can then be calculated internally. During calibration the controller sends a signal to indicate which input reference is being applied to the input of the sensor.

At the n-th calibration step, the reference signal x_n is applied by the calibration controller, corresponding to the desired output y_n. The input $f(x_n)$ is sampled, and processed through n-1 correction steps. The n-th step is then executed with a_n=0, and $h_{n-1}(x_n)$ is obtained at the output. The output signal is subtracted from the desired output y_n, and the result $\{y_n-h_{n-1}(x_n)\}$ is used to obtain the correct value for a_n. This can be achieved in two ways. As shown in the figure, the signal $\{y_n-h_{n-1}(x_n)\}$ can be *divided* by the value of the available correction term $\Pi\{h_i(x)-y_i\}$, in

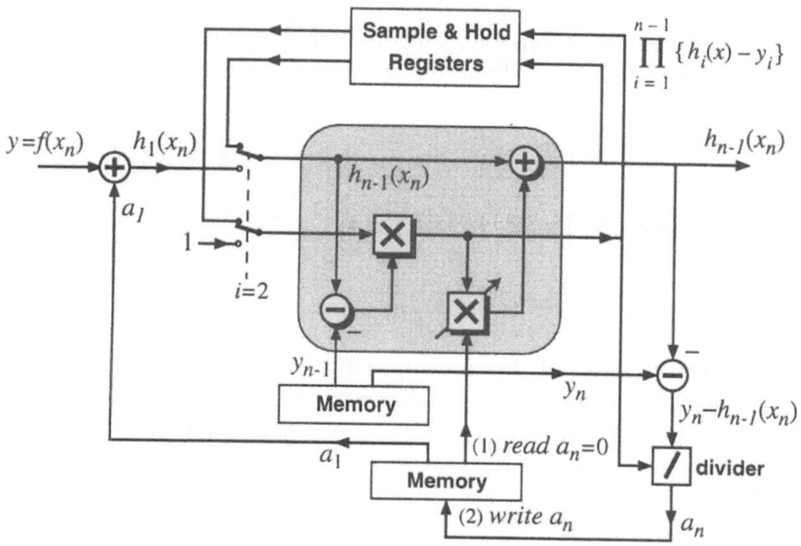

Fig. 6-2 *Algorithmic loop implementation with calculation of the calibration coefficient, at the n-th step of the calibration phase.*

accordance with the formula's in Table 6-1. The other way is to compute a_n algorithmically. As in a *Successive-Approximation* converter, the bits of a_n are determined, and programmed one by one from MSB to LSB, based on the polarity of the signal $\{y_n-h_{n-1}(x_n)\}$. Bit by bit that signal is reduced to zero, the output is equalized to the desired value, and the resulting value for a_n is stored in memory.

When considering implementation of the block diagrams in Fig.6-1 and Fig.6-2 in digital hardware, the following objection comes to mind. The digital implementation will turn out to be very similar to a standard microprocessor. Adders and multipliers are needed; these functions are usually efficiently implemented in the Arithmetic Logic Unit (ALU) of a microprocessor. Furthermore, the necessary memory addressing is also present in any microprocessor, and the sample & hold registers are available for example as 'stack' memory. The question rises whether the development of an custom processor for digital calibration would be worth the investment. The only advantages of a dedicated digital processor for the digital calibration would be the possibility to optimize for speed, power consumption or chip area, whereas flexibility and debugging options have to be missed. Hence the option to implement the progressive polynomial calibration method in a small software program for a standard microcontroller is preferred for experiments. It is presented in the next section.

6.3 Software implementation

6.3.1 Sensor system configuration

The purpose of the implementation of the two-dimensional calibration procedure in a microcontroller program [9, 10], is to eventually apply it in a smart sensor system set-up as proposed in Fig. 6-3.

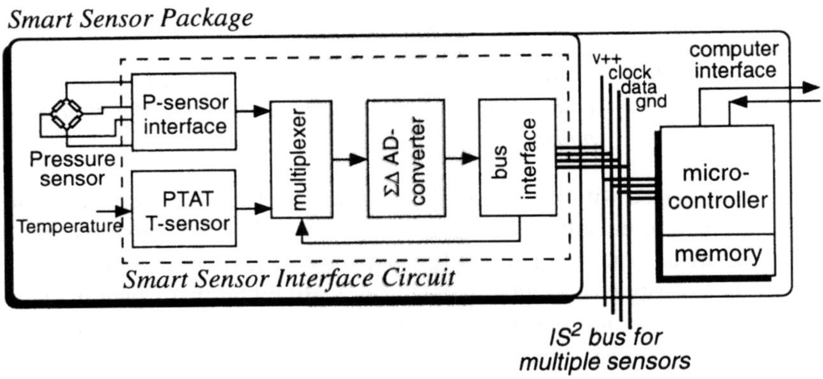

Fig. 6-3 *Pressure sensor system configuration for including a calibration function in microcontroller software.*

Using a smart sensor interface chip with $\mathbf{IS}^2$ bus interface the microcontroller can read out the pressure sensor and an integrated PTAT temperature sensor [11]. A multiplexer controlled by the bus interface makes it possible to read out both sensors using a single sigma-delta AD-converter. The output of the pressure sensor is determined by the applied pressure *and*, because of cross-sensitivity, to the operating temperature of the sensor. The sensor interface with PTAT sensor should thus be integrated close to the pressure sensor so that the measured temperature is related closely to that of the pressure sensor. Different levels of integration are possible, depending on which components can be integrated on the same chip, multi-chip module, or package. It is also possible to keep the microcontroller external, so it can be used for multiple sensors in a larger bus-wired system.

For the development of the microcontroller program it was assumed that the microcontroller can invoke samples of the pressure and temperature sensor[1], and that it must autonomously compute the corrected output pressure based on stored calibration data. Furthermore, it is desirable that the microcontroller program can autonomously compute the required calibration coefficients, when the sensor system is inserted in a calibration set-up. Through a computer interface the microcontroller must then be signalled about the reference inputs being applied and the desired output value.

The calibration program and the calibration coefficients must be stored in the memory of the microcontroller, preferably a non-volatile memory such as PROM.

6.3.2 Calibration algorithm

The two-dimensional polynomial calibration method has been explained shortly in section 3.3.2. The equations for an $N \times M$ calibration are adapted for a pressure sensor calibration, and summarized in Table 6-2. The computation of the calibration coefficients, using $N \times M$ calibration measurements $f(p_n, T_m)$ is presented in Table 6-3.

The pressure sensor output signal is represented by $y=f(p,T)$, whereas $h_{nm}(p,T)$ represent the subsequent calibrated sensor functions. The calibrated output signal is given by the last calibration function $h_{NM}(p,T)$. For the calibration of the sensor, N different pressure values have been used, corresponding to desired output values y_1 to y_N. At each pressure value, M different temperatures have been used for calibration. The signal T represents the output of the temperature sensor rather than the actual temperature. Also for the temperature calibration points T_m the temperature sensor outputs, measured in the calibration phase, must be used. In that way calibration of the errors of the temperature sensor will be included automatically.

---

[1] For this purpose the microcontroller needs to be equipped with an IS^2 bus interface, which can be realized either in hardware or in software. These developments have not yet been completed.

Table 6-2 *Overview of calibration functions for the two-dimensional progressive polynomial calibration method.*

n \ m	1	...	M
1	$h_{11}(p,T) = f(p,T) + a_{11}$	...	$h_{1M}(p,T) = h_{1,M-1}(p,T) + a_{1M} \cdot \prod_{m=1}^{M-1}(T-T_m)$
2	$h_{21}(p,T) = h_{1M}(p,T) + a_{21} \cdot \{h_{1M}(p,T) - y_1\}$	...	$h_{2M}(p,T) = h_{2,M-1}(p,T) + a_{1M} \cdot \{h_{1M}(p,T) - y_1\} \cdot \prod_{m=1}^{M-1}(T-T_m)$
...	...		...
N	$h_{N1}(p,T) = h_{N-1,M}(p,T) + a_{N1} \prod_{n=1}^{N-1} \{h_{nM}(p,T) - y_n\}$	...	$h_{NM}(p,T) = h_{N,M-1}(p,T) + a_{NM} \prod_{m=1}^{M-1}(T-T_m) \prod_{n=1}^{N-1} \{h_{nM}(p,T) - y_n\}$

Table 6-3 *Overview of calibration coefficients for the two-dimensional progressive polynomial calibration method.*

n \ m	**1** $T=T_1$	...	**M** $T=T_M$
1 $p=p_1$	$a_{11} = y_1 - f(p_1, T_1)$	...	$a_{1M} = \frac{y_1 - h_{1,M-1}(p_1, T_M)}{\prod_{m=1}^{M-1}(T_M - T_m)}$
2 $p=p_2$	$a_{21} = \frac{y_2 - h_{1M}(p_2, T_1)}{\{h_{1M}(p_2, T_1) - y_1\}}$	...	$a_{2M} = \frac{y_2 - h_{2,M-1}(p_2, T_M)}{\{h_{1M}(p_2, T_M) - y_1\} \cdot \prod_{m=1}^{M-1}(T_M - T_m)}$
...	...		...
N $p=p_N$	$a_{N1} = \frac{y_N - h_{N-1,M}(p_N, T_1)}{\prod_{n=1}^{N-1}\{h_{nM}(p_n, T_1) - y_n\}}$	...	$a_{NM} = \frac{y_N - h_{N,M-1}(p_N, T_M)}{\prod_{n=1}^{N-1}\{h_{nM}(p_n, T_M) - y_n\} \cdot \prod_{m=1}^{M-1}(T_M - T_m)}$

The arrows in Table 6-2 and Table 6-3 indicate the direction of execution. Along the rows of the tables the temperature dependency is calibrated, whereas the columns represent the calibration steps on the pressure axes. It is important that at each pressure calibration step the sensor curve is first corrected for temperature before proceeding to the next pressure calibration steps. Otherwise the temperature dependency gets multiplied by the pressure corrections, and gets harder to rectify. The example may clarifies this. It is assumed that generally the sensor signal depends more on pressure than on temperature.

The algorithm for the two-dimensional polynomial calibration has been tested, using the mathematical computer program MATLAB. This gives the obvious advantage that we can study the behavior of the calibrated sensor response, where in practice this would require an impractical number of measurements. For that purpose, several fictive sensor responses were generated, using mathematical descriptions of two-dimensional functions with different kinds of nonlinearity and cross-sensitivity. Fig.6-4 shows one such a uncalibrated sensor response used for testing. It clearly shows a slightly nonlinear pressure dependency and a small cross-sensitivity to temperature. The frame indicates the ideal transfer function.

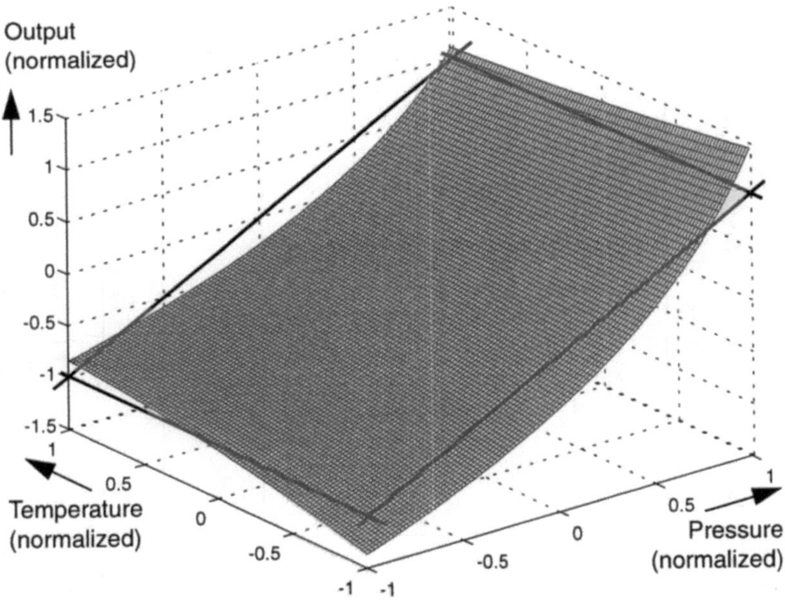

Fig. 6-4 *Sensor transfer function (example), output versus normalized pressure and temperature.*

The calibration is now started by measuring the sensor output at one pressure p_1, for different temperatures. After correction at for example three temperature points T_1, T_2, T_3, the output is more or less constant along the axis $p=p_1$, $y=y_1$. Fig.6-5 shows the sensor transfer in that situation.

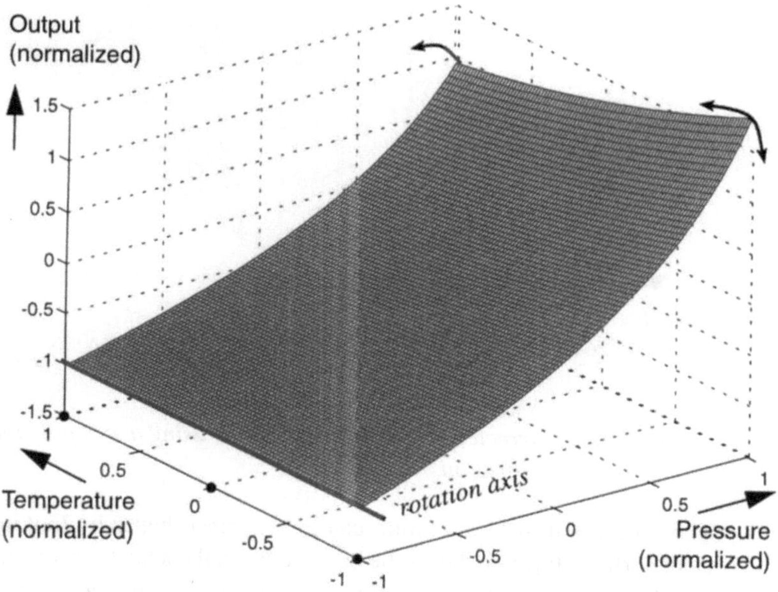

Fig. 6-5 *Sensor transfer curve, after calibration at one pressure point and three points along the temperature axis.*

It is now possible to rotate the response surface around that axis, using $\{h(p,T)-y_1\}$ for the gain correction. Because the error at $p=p_1$ is already made very small, it will not get affected much by the gain correction. At the next pressure calibration point p_2 the gain correction is determined. And again the temperature-dependency is reduced first by calibration at temperature points T_1, T_2, T_3, before proceeding to the next calibration step.

After 5×3 correction steps the transfer curve seems practically identical to the desired transfer curve, see Fig.6-6.

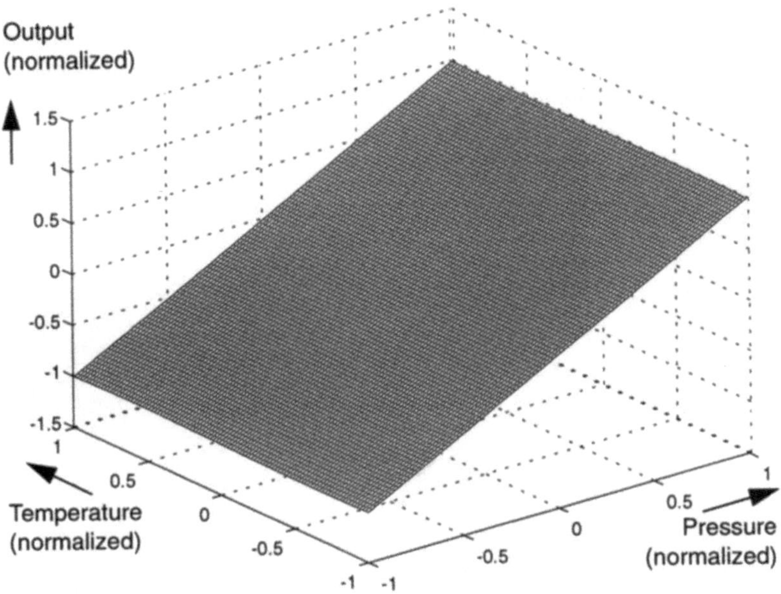

Fig. 6-6 *Corrected sensor transfer curve, using a 5×3 polynomial calibration.*

The progress in the calibration can be examined better by looking at the error surface. Fig. 6-7 shows the error between the actual sensor output and the desired output as a function of pressure and temperature, without calibration.

After correction based on a 5×3 calibration, the error surface is flattened as shown in Fig. 6-8. The error is made zero in all the used calibration points (p_1, T_1) to (p_5, T_3). Along the temperature and the pressure axes the third-order and fifth-order polynomial may be identified. The labeled points on the "ground" surface indicate the calibration order. After calibration the peak in the error surface is reduced to about 1 percent of the full-scale output.

Further reduction of the error may be obtained by using more calibration points. Fig. 6-5 shows the error surface in case the correction was based on a 7×5 polynomial calibration. The remaining error peak was reduced to a value smaller than 0.2%.

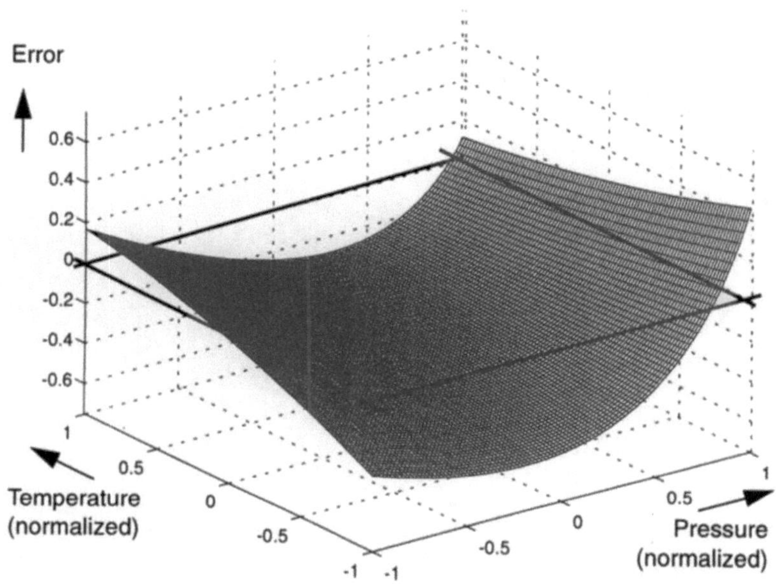

Fig. 6-7 *Initial error surface.*

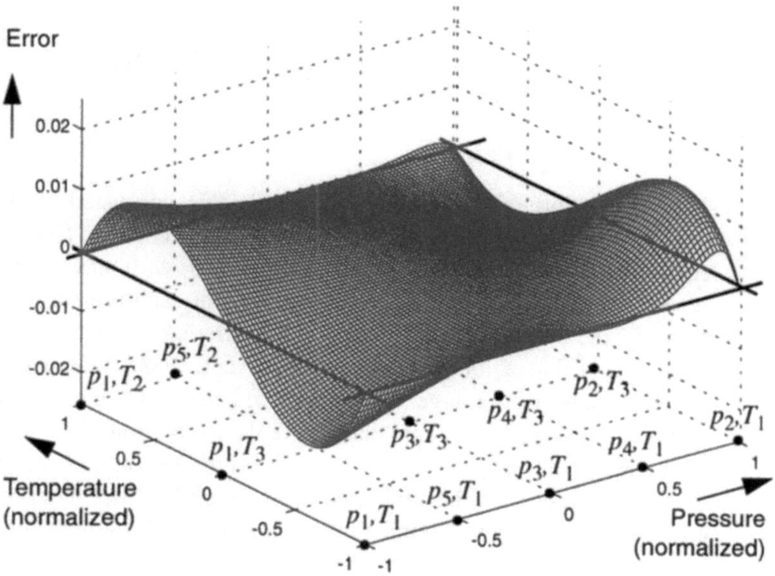

Fig. 6-8 *Error surface after correction based on a 5×3 polynomial calibration.*

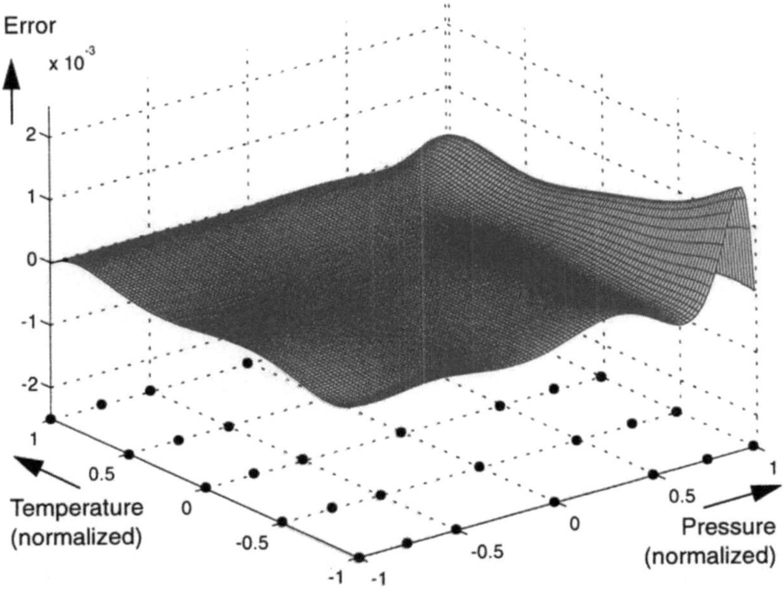

Fig. 6-9 *Error surface after correction based on a 7×5 polynomial calibration.*

The reduction of the error has been examined for several different sensor transfers, and different numbers of calibration steps.

Calibration coefficients

The calculation of the calibration coefficients suggested in Table 6-3, implies a certain order of the calibration measurements. The calibration coefficients in the table are calculated row by row. In case of a pressure sensor the order is disadvantageous, namely at each pressure calibration point p_n, the output must be measured for different temperatures T_1 to T_M. This means that a lot of temperature sweeps, $N \times M$, would have to be made. Because of the large settling time of a temperature chamber (oven), in which pressure sensors are calibrated, the calibration would take a long time and become expensive. For most pressure sensors, the test pressures for the calibration can be generated much quicker. Hence it is preferred to minimize the number of temperature steps, and perform multiple pressure measurements at each temperature.

Fortunately the calculation of the calibration can easily be modified to enable a different arrangement of the calibration measurements. Looking at the equations of the calibration functions, see Tables 6-2, we can observe that for $T=T_1$:

$$h_{1M}(p_2,T_1)=h_{11}(p_2,T_1).$$

Thus we don't have to compute the first row completely to be able to calculate the calibration coefficient a_{21} on the second row, as suggested in Tables 6-3. Because for all equations in the first column:

$$h_{n-1,M}(p_n,T_1)=h_{n-1,1}(p_n,T_1),$$

the calculation of the calibration coefficients can be modified as shown in Table 6-4. Now the coefficients can be calculated column by column as indicated with the arrows. Accordingly, the calibration measurements are done in that order as well. In that case the number of temperature sweeps is minimized to M, whereas the pressure needs to be changed $M \times N$ times.

In fact the calibration coefficients can be calculated in any order working through a similar table but always from left to right or downwards. In that way the calibration can also be expanded afterward, using extra columns and/or rows of calibration measurements.

Table 6-4 *Overview of calibration coefficients for the two-dimensional progressive polynomial calibration method.*

n \ m	1 $T=T_1$	…	M $T=T_M$
1 $p=p_1$	$a_{11} = y_1 - f(p_1, T_1)$	…	$a_{1M} = \dfrac{y_1 - h_{1,M-1}(p_1, T_M)}{\prod_{m=1}^{M-1}(T_M - T_m)}$
2 $p=p_2$	$a_{21} = \dfrac{y_2 - h_{11}(p_2, T_1)}{\{h_{11}(p_2, T_1) - y_1\}}$	…	$a_{2M} = \dfrac{y_2 - h_{2,M-1}(p_2, T_M)}{\{h_{1M}(p_2, T_M) - y_1\} \cdot \prod_{m=1}^{M-1}(T_M - T_m)}$
…	…		…
N $p=p_N$	$a_{N1} = \dfrac{y_N - h_{N-1,1}(p_N, T_1)}{\prod_{n=1}^{N-1}\{h_{n1}(p_n, T_1) - y_n\}}$	…	$a_{NM} = \dfrac{y_N - h_{N,M-1}(p_N, T_M)}{\prod_{n=1}^{N-1}\{h_{nM}(p_n, T_M) - y_n\} \cdot \prod_{m=1}^{M-1}(T_M - T_m)}$

6.3.3 Calibration program

The microcontroller we used for evaluation of the calibration method is the 80c552 which is an inexpensive 8-bit microcontroller driven by a 12 MHz clock. The system software consists of a program written in C, which is compiled, assembled, and then downloaded into the microcontroller memory through an RS232 interface. And the program could be tested on microcontroller evaluation board, using PC-communication.

The program consists of two parts. First the *calibration part,* to calculate the calibration coefficients when applying reference signals to the sensors. At each calibration step the output signals are measured, and processed according to the equations in Table 6-2, to compute the calibration coefficients as described in Table 6-4. Second the *measurement or correction part,* to calculate the corrected values of the pressure output, according to equations in Table 6-2, using the sensor outputs and the values of calibration coefficients already calculated during the calibration cycle. From the implementations of the one-dimensional polynomial calibration it was already clear that both the calibration and the correction can be implemented algorithmically. In programming this means that the calculations are based on the repetitive use of the same procedure. For the two dimensional calibration a nested loop is needed. Altogether it results in a very compact program. The complete calibration program, together with the calibration coefficients and reference data for a 5×5 calibration, required only 6.6 kilobytes memory of the microcontroller.

Initially the program is developed as an interactive program which during the calibration process inquires the user about the reference data and the desired output signal, at each stage of the calibration process. Then the microcontroller proceeds with the necessary calculations, which involves calculating a new calibration coefficient and checking the error limit supplied by the user. Once this limit has been reached, entering more calibration data can stop, and the measurement cycle can start. Once the optimal number of calibration points and reference signals has been chosen, the system can be made autonomous by storing the program and the reference data in the microcontroller PROM, and by using hardware interrupts to indicate that certain reference signals (pressure and temperature) have been applied for calibration. During batch calibration, many of such pressure sensor systems can then be connected to the same interrupt signals, and each system independently calculates and stores its own calibration coefficients.

The microcontroller program was operated and tested, using the fictive sensor data already used in the simulations. As expected, the actual results from the microcontroller program were found to coincide with the simulation results. With the microcontroller type already mentioned, and 32-bit precision floating point arithmetic, it was found that the time required to calculate one corrected value during the measurement cycle was approximately 30 ms based on 5×5 calibration steps. This is fast enough for sensors with a signal bandwidth lower than 15 Hz. Of course speed and accuracy can be exchanged. To increase speed one could for example reduce calibration accuracy to 5×3 or 3×3 steps, chose a lower arithmetical resolution, or chose a faster 16-bit (more expensive) microcontroller. Maximum speed can then be in the order of 100-1000 samples per second.

Finally, Fig.6-10 gives an overview of the remaining peak errors after several calibration steps, for six different sensor curves that were used for testing of the two-dimensional calibration method.

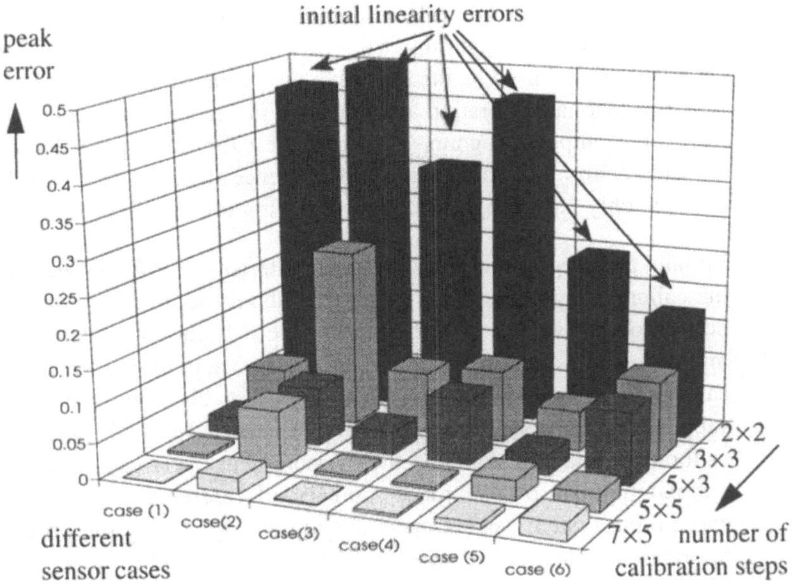

Fig. 6-10 *Overview of the remaining peak errors for different sensor cases, when increasing the number of the calibration steps.*

For the six test cases, different pressure and temperature dependencies have been combined, using polynomial, root, exponential, and logarithmic functions. The large bars represent the initial linearity error, since with a 2×2 calibration only the linear errors are corrected. The figure shows how the linearity error reduces when increasing the number of calibration measurements to respectively 3×3, 5×3, 5×5, and 7×5. One has to keep in mind that the bars indicate the maxima of the error surfaces

6.3.4 Future work

Still some work remains to be done before the calibration program can actually be applied in the intended smart sensor configuration of Fig.6-3 and tested with real sensor data. The developed pressure sensor interface described in section 2.3.2 needs to be combined on one chip with a smart temperature sensor as described in section 5.5.1. The microcontroller program must be expanded with a bus-interface for sensor read-out. Also the packaging for the combination of the sensor, the interface IC, and the microcontroller must be investigated. Further investments are required to set up a (computer-) controlled measurement configuration for pressure and temperature, which is needed to calibrate and test such smart sensor.

6.4 Conclusion

In the consideration of digital implementations of calibration techniques we concentrated on the proposed polynomial calibration method. The step-by-step correction of the sensor transfer curve can be set up in an algorithmic way. For each calibration step the same calculation procedure is then used. This applies for the correction of the sensor transfer function as well as for the computation of the calibration coefficients. It allows a compact realization in hardware in the form of an algorithmic loop implementation. When implemented in software, the required calibration program can also be made very compact. The latter option can be successfully applied in future microcontroller-based smart sensors. With respect to the implementations of the previous two chapters, it offers the advantage of an advanced two-dimensional calibration, at the costs of (hybrid) integration of a microcontroller, and some sacrifice in sensor signal bandwidth.

References

[1] J.E. Brignell, "Software techniques for sensor compensation", *Sensors and Actuators A*, Vol.25-27, 1991, pp.29-35.

[2] D. Patranabis and D. Gosh, "A novel software-based transducer linearizer", *IEEE Trans. Instrumentation and Measurement*, Vol.36, No.6, December 1989.

[3] S.B. Crary, W.G. Baer, J.C. Cowles, and K.D. Wise, "Digital compensation of high performance silicon pressure transducers", *Sensors and Actuators A*, Vol.21-23, 1990, pp.70-72.

[4] S. Huang, R.Z. Morawski, and A. Barwicz, "Static calibration based on superposition of splines in one variable", *Proceedings IMTC'96*, June 1996, pp.49-53.

[5] W.T. Bolk, "A general digital linearising method for transducers", *Journal of Physics*, Vol. E: Scientific Instrumentation, 1985, pp.61-64.

[6] P.N. Mahana and F.N. Trofimenkoff, "Transducer output signal processing using an eight-bit microcomputer", *IEEE Trans. Instrumentation and Measurement*, Vol.IM-35, No.2, June 1986, pp.182-186.

[7] S.R. Ashley, M. Muggeridge, and J. Lucas, "An inexpensive digital linearizer for nonlinear transducers", Journal of Physics E: Sci. Instrum., Vol.11, 1978, pp.576-580.

[8] C. Berthoud, M. Ansorge, and F. Pellandini, "Effective static response compensation suitable for low-power ASIC implementation with an application to pressure sensors", *Proceedings IMTC'96*, June 1996, pp.1168-1173.

[9] K.F. Lyahou, G. v.d. Horn, and J.H. Huijsing, "A non-iterative polynomial 2-dimensional calibration method implemented in a microcontroller", *Proceedings IMTC'96*, June 1996, pp.62-67.

[10] G. v.d. Horn, K.F. Lyahou, and J.H. Huijsing, "Calibration and linearization method for microcontroller-based sensor systems", *Journal on Measurement + Control*, Vol.29, No.9, November 1996.

[11] G. v.d. Horn and J.H. Huijsing, "Integrated smart sensor calibration", *J. Integrated Analog Circuits and Signal Processing*, Vol.14, No.3, November 1997.

Summary and Conclusions 7

Smart Sensors

Electronic sensors are applied in industrial machines, as well as in consumer products, in order to enhance functionality, or to improve performance and reliability. Currently, the application of sensors requires a high-level expertise, due to the large diversity in sensors and, correspondingly, in sensor output formats, and in the required electronic sensor interfaces. The application of electronic sensors will be greatly stimulated if the sensors are equipped with smart electronics, which provide conversion of the sensor signal to a standardized digital output and bus interface. Using modern silicon IC-technology, such smart sensors can be produced in high volume and at low cost. This will make them available for application in more products, and may also lead to new products. A smart sensor configuration, composed of silicon sensor(s), analog front-end circuit, one-bit sigma-delta AD-converter, and a serial bus interface, forms a suitable concept for complete integration.

Calibration

For standardization, a fixed transfer from sensor input to electrical output is desired for each sensor (type). Due to variations in the production, most sensors exhibit some unpredictable errors with respect to the desired transfer. Each sensor needs to be calibrated individually to accomplish the desired fixed transfer. This has to be realized in an inexpensive way, which means that the calibration procedure has to be automated in the sensor production line. Currently, sensor calibration requires an expensive (time consuming) laser-trimming technique. Automated calibration could become feasible if electronics are included in the smart sensor, to enable correction of the sensor transfer function in a digitally programmable (computer-controlled) way. Furthermore, the costs of calibration in a sensor production line can be reduced by executing a minimum number of calibration measurements, and by measuring a maximum number of sensors at a time. The latter is facilitated if the smart sensors are equipped with a bus interface. The first must be obtained by choosing an effective linearization method.

Linearization

Different linearization methods have been examined. Generally, the use of a single (full-range) interpolation function, or a set of (sub-range) interpolation functions (splines), is the best choice for linearizing the sensor transfer curve, using a minimum number of measurement data. Classically, the sensor transfer curve is first measured completely (all calibration points), and then all the required calibration coefficients are calculated. The progressive polynomial calibration method is proposed, which linearizes the sensor transfer curve in a step-by-step manner. At each step, one calibration measurement is done, and one calibration coefficient is calculated and applied immediately. Every succeeding calibration step is done in such a way that the previous calibration(s) remain undisturbed. At each step, the calibrated transfer curve progresses towards the desired transfer curve. In addition to the linearization of a one-dimensional sensor transfer function, the progressive method is also worked out for the two-dimensional polynomial calibration of sensors, showing a cross-sensitivity error.

Implementation

At several places in the signal processing chain of a smart sensor, the overall transfer function can be modified. Traditionally, the sensor transfer function is adjusted in the analog (physical) signal domain, by means of trimmable components, or in the digital domain, by using a digital memory with a look-up table, or by using a data-acquisition computer. These three options are problematic concerning the desired features for the calibration function, respectively, programmability, a low number of calibration points, and integration with the sensor.

It was shown that an advanced two-dimensional calibration procedure, based on the progressive polynomial linearization method, can be implemented in a simple microcontroller program. The calibration program has been tested with computer-generated fictive sensor data, for different numbers of calibration steps, up to 7×7 measurement points. This (software) calibration feature can be added to one sensor exclusively, by hybrid integration of a microcontroller die and a sensor die in one package. After the packaging, an automated sensor calibration procedure can be inserted in the production line in order to manufacture calibrated smart sensors. At the current technology level, it is often more cost-effective to use one microcontroller, in a complete (bus-wired) data-acquistion system, for readout and calibration of multiple sensors. However, this implies that the sensors will still have to be calibrated by the customer who applies the sensors in the data-acquistion system. Alternative options have therefore been examined.

In the analog signal path, the sensor transfer function can be altered in a digitally programmable way, by using Digital-to-Analog (DA) converters. Calibration of offset, gain, and cross-sensitivity errors can be implemented in a classical approach. These implementations can be modified in accordance with the progressive calibration method, to eliminate influence between successive settings (offset, gain, etc.), and to circumvent iteration of the adjustments. A programmable analog circuit is presented which implements four steps of the progressive polynomial calibration method. Programmable analog circuits can be used for sensor calibration with limited resolution (8-10 bit) and limited correction of higher-order effects (nonlinearity/cross-sensitivity). As an advantage, a large bandwidth can be maintained for the (analog) transfer function, as it is not limited by any digital sampling or processing speed.

As an alternative for the DA-conversions, digital pulse modulation can be used to modify the transfer of an analog signal. This technique has been elaborated in combination with sigma-delta AD-conversion. The technique multiplies analog input signals with digitally-controlled bitstream (pulsed) signals. This offers the possibility of a high-resolution correction. Because the signals are all summed at one point, the technique should be applied only to calibration of (lower-order) non-idealities that can be compensated with additive correction signals. The use of the bitstream modulation technique is demonstrated by the realization of two smart temperature sensor configurations, providing calibration of offset, gain, and second-order linearity errors.

To summarize, the Table 7-1 rates the important features of the examined implementations of a programmable calibration function.

Table 7-1 *Ratings for the features of the different implementations of a programmable sensor calibration function.*

Calibration Features	**Analog Signal Processing**	**Bitstream Modulation, AD-conversion**	**Digital Signal Processing**
Offset, Gain Cal.	+++	+++	+++
Linearity Calibration	+	+	+++
Cross-sensitivity Cal.	+	+	+++
Progressive Polynomial Calibration	+	--	+++
Resolution	-	++	++
Bandwidth	++	+	-
Integration with Sensor (interface)	++	+++	--
Flexibility	--	--	+++

---=impossible/very low,...,+++=very suitable/very high

For low-resolution calibration of a small number of error corrections, the use of an analog circuit with programmable DA-converters is feasible. It is suitable for integration with the sensor or sensor interface, provided that a high-density (VLSI) integration technique, compatible with the sensor technology, is available. If the advantages of the analog circuit technique, namely, a compact integration (relatively low number of transistors) and large 'analog' bandwidth, are not really needed, then a digitally oriented calibration technique is favored.

For a low number of high-resolution corrections, the bitstream modulation technique is the best option. It is also a suitable technique for integration of the calibration circuit in the proposed smart sensor concept. Because of the amount of digital circuitry required for the generation of the bitstream signals, a CMOS implementation would be more compact and thus preferable.

For high-resolution calibration with an advanced correction method (such as for higher-order polynomial linearization and cross-sensitivity compensation), the use of a microcontroller or a dedicated digital signal processor is the best option. Advances will still have to be made to make the integration of such digital signal processors with sensors economically feasible.

Several techniques have been presented which make it possible to include a digitally programmable calibration function in a sensor interface circuit. To become useful, these techniques must be combined with (existing) non-volatile memory techniques for storage of the digital calibration coefficients. A calibration circuit can be integrated with each sensor at an increase in the production costs, which is related to the silicon surface of the total calibration circuit. However, by making use of this enhanced functionality of the sensor for automation of the calibration procedure, the production costs of accurate (factory-calibrated) sensors can be reduced. Some costs per calibration measurement will remain (heating a temperature chamber costs time). For some kinds of sensors the netto effect will be a reduction of production costs per sensor. Considering the rapidly increasing integration densities and decreasing price per transistor, an integrated smart sensor with calibration feature will become more and more viable in the future. Furthermore, the calibration function that can be integrated with a sensor will become more advanced (faster, more accurate, or a higher-order linearization). As usual in circuit technology, and also for the calibration circuits, accuracy, speed, and circuit complexity can be exchanged.

Finally, it should be recognized that calibration techniques can only improve some aspects of a sensor, and at a relatively high price (calibration measurements). If a sensor transfer function can be improved in another way, for example by using an improved construction or material or an auto-calibration, this is given preference. Research on such sensor developments deserves to be stimulated.

Index

A

actuators 2
AD conversion 5, 33
AD converter 33
  sigma-delta converter 127
analog calibration
  cross-sensitivity 99
  full-scale or gain 98
  IC realization 121
  implementation 106
  independent 103
  offset 98
  pulse-modulation 104
  temperature coefficients 101
analog signal processing 93

B

bandgap reference 27
bandgap voltage 23
batch calibration 93
bitstream 36, 104
bitstream generator 140
bitstream spectrum 151
bus architecture 39
bus interface 5
  IS2 interface 43
bus master 40
bus protocol 40
  IS2 protocol 42
bus slave 39

C

calibration 5, 7
  analog implementation 106
  auto-calibration 13
  batch calibration 7
  definition 8
  errors 120
calibration algorithm 179
calibration coefficient 78, 186
calibration memory 46
calibration procedure 118, 167, 189
calibration program 189
cascode dividers 111
charge balance 38
comparator 34
  implementation 155
controller 3
cross-sensitivity 57
curve-fitting 70

D

DA conversion 96
DA converter
  implementation 111
  multiplying DAC 97, 111
DAC coefficient 103
data-acquisition 3
decimation 136, 138
decimation filter 136
  implementation 159

digital calibration
  hardware implementation 174
  one-dimensional 174
  progressive polynomial 175
  software implementation 178
  two-dimensional 179

discrete-time domain 38

dithering 140, 145, 150

downsampling 128, 136

downsampling ratio (DSR) 128

drift 8

E

Emitter-Coupled Logic 116

error compensation 54

error minimization 73

error surface
  two dimensional 184

errors
  cross-sensitivity 52
  distribution 54
  drift 52
  error distribution 25
  full-scale 52
  gain 52
  hysteresis 52
  linear versus nonlinear 55
  measurement errors 6, 25
  nonlinearity 52
  offset 52
  sensor characteristic 53
  systematic versus random 53

F

feedback testing 15

feedforward testing 15

full adder .141

I

in-situ calibration 93

integrator 34
  implementation 153
  transfer function 131

interpolation
  Lagrange 71, 80
  piece-wise linear 63
  splines 69

interpolation curve 65

IV converter 114

L

Lagrange interpolation 71, 80

laser trimming 94

limit cycles 142

linearization 55
  curve fitting 70
  look-up table 60
  piece-wise linear 63
  splines 67

linearization methods 56, 75

look-up table 60

M

Manchester coding 41

microcontroller 5
  interface 39
  programs 173

micromachining 23

mismatch effect 31

multiplier
  Gilbert multiplier 101
  implementation 108, 165

N

noise shaping 128, 130, 136
- first and second order 133
- spectrum 135

normalized input signal 56

normalized output signal 56

normalized transfer function 56

Nyquist rate sampling 130

O

oversampling 128- 129

oversampling ratio (OSR) 36, 130

P

piece-wise polynomial interpolation 67

piecewise-linear interpolation 63

piezo-resistors 24

pressure sensor 23
- readout circuit 32

progressive polynomial calibration . 77
- algorithmic loop 176
- one-dimensional 77
- signal flow diagram 84
- software implementation 178
- successive approximation . . . 177
- two-dimensional 84

pseudo-random noise 150

PTAT voltage 23

pulse stream 104

pulse-density modulation 104

pulse-width modulation 104

Q

quantization 129

quantization level 61

quantization noise 131

quantizer 131

R

ratiometric conversion 32

resistor arrays 96

S

self-calibration 13

self-test 11

sensor 2
- pressure sensor 23
- temperature sensor 22

sensor transfer function
- normalized transfer function . . . 56
- one-dimensional 56
- two-dimensional 58, 183

sigma-delta 33
- bitstream generator 141
- principle 34
- signal multiplication 149

sigma-delta AD converter 128
- differential 128
- discrete-time model 131
- implementation 153, 157
- input-output transfer 139
- output spectrum 135
- signal processing chain 129
- signal-to-noise ratio . 130, 132- 134

sigma-delta calibration
- cross-sensitivity 148
- gain 143
- modulated spectra 147
- offset 147

sigma-delta DA-converter 140

signal conditioning 5

signal flow diagram 106

signal transfer
- correction 45
- errors 52

signal transfer function 56
inverse function 70
logarithmic 59
one-dimensional 56
two-dimensional 57

sinc filter 136

sinc2 filter .137

smart sensor .4
calibrated temp. sensor . . .161, 165
configuration 43
integrated smart sensor 5
interface 21
interface chip 178
IS2 bus 40
microcontroller configuration . 178
signal chain 44
smart temperature sensor 29

splines 67

standard unit 6

T

temperature sensor 22
implementation 160, 165

translinear circuit 108

trimming 94

two's complement 112, 142

U

user interface 3

V

VI converter 113

W

Wheatstone bridge 25

window function 136
triangular window 137
uniform window 136

MIX
Papier aus verantwortungsvollen Quellen
Paper from responsible sources
FSC® C105338
FSC
www.fsc.org

If you have any concerns about our products,
you can contact us on
ProductSafety@springernature.com

In case Publisher is established outside the EU,
the EU authorized representative is:
Springer Nature Customer Service Center GmbH
Europaplatz 3, 69115 Heidelberg, Germany

Printed by Libri Plureos GmbH
in Hamburg, Germany